Essential Maths Skills
for AS/A-level
Physics

Ian Lovat

PHILIP ALLAN FOR
HODDER EDUCATION
AN HACHETTE UK COMPANY

Philip Allan, an imprint of Hodder Education, an Hachette UK company, Blenheim Court, George Street, Banbury, Oxfordshire OX16 5BH

Orders

Bookpoint Ltd, 130 Park Drive, Milton Park, Abingdon, Oxfordshire OX14 4SE

tel: 01235 827827

fax: 01235 400401

e-mail: education@bookpoint.co.uk

Lines are open 9.00 a.m.–5.00 p.m., Monday to Saturday, with a 24-hour message answering service. You can also order through the Hodder Education website: www.hoddereducation.co.uk

© Ian Lovat 2016

ISBN 978-1-4718-6343-1

First printed 2016

Impression number 5 4 3 2 1

Year 2020 2019 2018 2017 2016

All rights reserved; no part of this publication may be reproduced, stored in a retrieval system, or transmitted, in any other form or by any means, electronic, mechanical, photocopying, recording or otherwise without either the prior written permission of Hodder Education or a licence permitting restricted copying in the United Kingdom issued by the Copyright Licensing Agency Ltd, Saffron House, 6–10 Kirby Street, London EC1N 8TS.

Typeset in India

Cover illustration: Barking Dog Art

Printed in Spain

Hachette UK's policy is to use papers that are natural, renewable and recyclable products and made from wood grown in sustainable forests. The logging and manufacturing processes are expected to conform to the environmental regulations of the country of origin.

Contents

The content listed in bold is only specified to be assessed at A-level.

Introduction 5

1 Units, standard form and orders of magnitude

Powers of 10 7
Units 10
Prefixes 11
Converting units 13
Symbols 16
Standard form and significant figures 17
Orders of magnitude 24

2 Fractions, ratios and percentages

Fractions 27
Ratios 31
Percentages 32

3 Averages and probability

Averages 35
Probability 36

4 Algebra

Re-arranging equations 39
Evaluating equations 42
Quadratic equations 44

5 Graphs
Straight lines 48
Shapes of graphs for different functions 50
Rates of change 55
Area under a graph 58

6 Geometry and trigonometry
Radians 64
Degrees and radians 65
Sine, cosine and tangent 67
Small angles 69
Pythagoras 73
Resolving vectors 75
Areas and volumes of simple shapes 79

7 Exponential changes
Radioactive decay 83
Capacitor discharge 87

8 Logarithms
Understanding logarithms 90
Logarithmic scales 91
Using logarithms 95
Logarithms to different bases 98

9 Uncertainty
Calculating uncertainty 101

Exam-style questions 109

Appendix
Specification cross-reference 118

Full worked solutions at www.hoddereducation.co.uk/essentialmathsanswers

Introduction

As well as introducing a lot of exciting new ideas, the successful study of physics at A-level also involves a wide range of mathematical skills. Physics is a numerical subject and you need to be familiar with, and able to use competently, a wide range of mathematical skills. In some cases these skills are those you have already encountered at GCSE but some will be new to you. This book covers the mathematics required for the A-level specifications set by AQA, OCR, Edexcel, WJEC/Eduqas and CCEA and you can use it to help you understand, develop and practise the ideas and techniques you will need.

When you come across a mathematical idea or technique that you are unsure of, you should check it in the book and use the questions provided to practise the skills involved. The best way to improve is to attempt lots of questions. The worked examples and questions in this book involve physics topics from across the course so that you can see how to apply the processes in different situations.

Understanding units and quantities, being able to analyse practical work and the ability to re-arrange and evaluate equations are skills that are fundamental to the study of physics and so the first sections of the book will be immediately relevant to you whatever course you are doing and whichever topic you start with. If you are at all uncertain about mathematics, you should tackle these sections first and then use the remaining sections as the need arises. There are examination-style questions at the end of the book for you to test yourself.

Calculus is not specifically required for any specification but a brief explanation of the symbols and ideas is given so that, if your teacher or textbook refer to calculus, you can follow the ideas and the meaning.

Full worked solutions to the guided and practice questions and exam-style questions can be found online at www.hoddereducation.co.uk/essentialmathsanswers.

1 Units, standard form and orders of magnitude

In order to complete an A-level Physics course successfully, you will need to be familiar with, and able to use confidently, a variety of mathematical techniques. These will all be covered in this book. However, the first and probably the most important requirement is that you can deal confidently with units, prefixes and standard form.

Powers of 10

Before dealing with prefixes, we need to be clear about powers of 10 and the notation used. $100 = 10 \times 10 = 1.0 \times 10^2$, often written just as 10^2. $1000 = 10 \times 10 \times 10 = 1.0 \times 10^3$, often written just as 10^3. Bigger numbers are written in a similar way. This is summarised in Table 1.1. Powers of 10 are used when numbers are expressed in 'standard form', which is covered later in this section.

Table 1.1 Positive powers of 10

Number	Written	Often written
10	1.0×10^1	10
100	1.0×10^2	10^2
1 000	1.0×10^3	10^3
10 000	1.0×10^4	10^4
100 000	1.0×10^5	10^5
1 000 000	1.0×10^6	10^6

By convention, above 1000, groups of digits are separated in groups of three by a space, for example, 100 000. Although commonly used, you should avoid using a comma to separate groups of digits. There is a good reason for this: outside the UK and the USA, a comma is often used as a decimal point and mistaking the position of a decimal point when administering drugs or designing a bridge could be disastrous.

Using a similar technique, $\frac{1}{10} = 1.0 \times 10^{-1}$, often written just as 10^{-1}. $\frac{1}{100} = 1.0 \times 10^{-2}$, often just written as 10^{-2}. These are summarised in Table 1.2.

Table 1.2 Negative powers of 10

Fraction	Decimal	Standard form	Often written
$\frac{1}{10}$	0.1	1.0×10^{-1}	10^{-1}
$\frac{1}{100}$	0.01	1.0×10^{-2}	10^{-2}
$\frac{1}{1000}$	0.001	1.0×10^{-3}	10^{-3}
$\frac{1}{10000}$	0.0001	1.0×10^{-4}	10^{-4}
$\frac{1}{100000}$	0.00001	1.0×10^{-5}	10^{-5}
$\frac{1}{1000000}$	0.000001	1.0×10^{-6}	10^{-6}

There are a couple of particular cases that you should know: $10^1 = 10$ and $10^0 = 1$. In fact, $x^0 = 1$ where x is any number.

You also need to remember that, for example, $\frac{1}{10^{-2}} = \frac{1}{1/100} = 100 = 1 \times 10^2$ so dividing by 10^{-x} is the same as multiplying by 10^x.

Multiplying and dividing powers of 10

In A-level Physics you will often be working with either very large numbers or very small numbers which will be written as a power of 10 and so it is important you understand how to multiply or divide powers of 10.

$1000 \times 100 = 100\,000$. Put another way, $10^3 \times 10^2 = 10^5$. In multiplying the numbers, we have added the indices, $3 + 2 = 5$.

Dividing the two numbers, $\frac{1000}{100} = 10$ or, put another way, $\frac{10^3}{10^2} = 10^1$. Here we have subtracted the power or index of the denominator, 2, from the power of the numerator, 3, so that $3 - 2 = 1$, in order to carry out the division calculation.

If you have a calculation such as $\frac{10^3 \times 10^5}{10^6}$, first add the indices in the numerator, $3 + 5 = 8$, then subtract the power of the denominator, $8 - 6 = 2$. Thus $\frac{10^3 \times 10^5}{10^6} = 10^2$.

> **TIP**
>
> When entering powers of 10 into a calculator, for example 1×10^2, students often enter 10×10^2 by mistake. You should always enter 1×10^x by entering '1' then pressing the button marked '$\times 10^x$' then entering the power of 10. On some calculators the '$\times 10^x$' button is labelled 'exp'.

You may frequently be presented with a calculation that involves two indices, for example, the area, A, of a circle of radius $r = 1.0 \times 10^{-3}$ m, is given by $A = \pi r^2 = \pi \times (1.0 \times 10^{-3})^2 = 3.14 \times 10^{-6}$ m^2. Here, $(1.0 \times 10^{-3})^2$ is equal to $1.0 \times 10^{-3} \times 1.0 \times 10^{-3}$ and so we add the two indices, $-3 + (-3) = -6$.

A Worked examples

a Write 1 000 000 as a power of 10.

Step 1: count the number of zeros after the 1.

There are six zeros, which is the same as multiplying 1 by 10^6.

Step 2: write the answer as a power of 10.

$1\,000\,000 = 10^6$

b Write 0.001 as a power of 10.

Step 1: write the number as a fraction.

$0.001 = \frac{1}{1000} = \frac{1}{10^3}$

Step 2: write the fraction as a power of 10.

$0.001 = \frac{1}{10^3} = 10^{-3}$

c Without using a calculator, evaluate the following:

i $10^3 \times 10^2$

Step 1: when multiplying, add the indices.

$3 + 2 = 5$

Step 2: write the answer as a power of 10.

$10^3 \times 10^2 = 10^5$

ii $10^8 \div 10^6$

Step 1: when dividing, subtract the denominator (bottom) index from the numerator (top) index.

$8 - 6 = 2$

Step 2: write the answer as a power of 10.

$10^8 \div 10^6 = 10^2$

iii $3.0 \times 10^6 \times 2.0 \times 10^{-4}$

Step 1: multiply the numbers 3.0×2.0 to give 6.0.

Step 2: add the indices.

$6 + (-4) = 2$

Step 3: put the two together.

$3.0 \times 10^6 \times 2.0 \times 10^{-4} = 6.0 \times 10^2$

B Guided questions

Copy out the workings and complete the answers on a separate piece of paper.

1 Write 0.00001 as a power of 10.

Step 1: write the number as a fraction.

$0.00001 = \dfrac{1}{100\,000} = \dfrac{1}{10^5}$

Step 2: write the fraction as a power of 10.

$0.00001 = $ _____

2 Write 10 000 000 as a power of 10.

Step 1: count the number of zeros after the 1.

There are seven zeros, which is the same as multiplying by 10^7.

Step 2: write the answer as a power of 10.

$10\,000\,000 = $ _____

3 Venus orbits about 100 000 000 km from the Sun.

Write this distance as a power of 10.

Step 1: count the number of zeros after the 1.

There are eight zeros, which is the same as multiplying by 10^8.

Step 2: write the answer as a power of 10.

$100\,000\,000 \text{ km} = $ _____ km

4 **Without using a calculator, carry out the following calculations:**

   a $1.0 \times 10^4 \times 1.0 \times 10^5$

   Step 1: when multiplying, add the indices.

   Step 2: write the answer as a power of 10: $1.0 \times 10^{\text{sum obtained}}$.

   b $1.0 \times 10^{12} \times 1.0 \times 10^9$

   Step 1: when multiplying, add the indices.

   Step 2: write the answer as a power of 10: $1.0 \times 10^{\text{sum obtained}}$.

   c $1.0 \times 10^9 \times 1.0 \times 10^{-5}$

   Step 1: when multiplying, add the indices. Even though one of the indices is negative, they are still added, but as negative numbers.

   Step 2: write the answer as a power of 10: $1.0 \times 10^{\text{sum obtained}}$.

   d $1.5 \times 10^3 \times 2.0 \times 10^{-6}$

   Step 1: multiply the numbers 1.5 and 2.0 together first.

   Step 2: add the indices to give the power of 10.

   Step 3: put the two together: 1.5×2.0 then $\times 10^{\text{sum obtained}}$.

   e $1.0 \times 10^9 \div 1.0 \times 10^6$

   Step 1: when dividing, subtract the denominator (bottom) index from the numerator (top) index.

   Step 2: write the answer as a power of 10: $1.0 \times 10^{\text{sum obtained}}$.

Practice questions

5 Your calculator shows 0.000000001 as an answer to a calculation. Write this as a power of 10.

6 A news article tells you that a star is one hundred million million km away. Give this distance in km as a power of 10.

7 In 2013 the processor chip for mobile phones had about one billion transistors on it. Give this as a power of 10.

8 Calculate $\dfrac{1.8 \times 10^{-8} \times 10}{1.0 \times 10^{-6}}$ without using your calculator.

Units

Almost every quantity you will encounter in your A-level Physics has a unit associated with it although there are a few, for example, tensile strain or refractive index, that are ratios and so do not have a unit. It is not necessary to put a unit after every figure in a calculation or after every step but you should always check at the end whether your answer should have a unit or not and make sure you put the correct unit after the final answer.

> **TIP**
>
> Always check your answer for a unit. If it should have a unit and you have not included it, the answer is wrong.

At GCSE you will probably have used the format m/s, for example, to mean metres per second. At A-level and beyond, the format m s^{-1} is used instead. If we take the GCSE format of m/s, this is $\frac{m}{s}$ = m × $\frac{1}{s}$. However, $\frac{1}{s}$ can also be written s^{-1} in the same way as $\frac{1}{10}$ can be written as 10^{-1}. The unit for acceleration is m/s^2 (metres per second per second) and at A-level this is written m s^{-2}.

This makes it easier to see exactly what the units are when they are combined. For example, the unit for specific heat capacity is J kg^{-1} K^{-1}, which could be confusing if not written in this format. If describing this to someone, you would say 'joules per kilogram per Kelvin'.

Prefixes

In the SI system of units there is a set of prefixes and when these are used, the power of 10 notation is avoided. You will be familiar with many of these already: for example, kilometre, or km, is 1000 m. A prefix, as the name implies, always goes before the unit name itself.

The danger with using these is that they can be easy to miss when they are used in the data for a question and the conversion can be forgotten. Always check the question carefully as this is a common source of error in calculations.

At A-level the commonly used prefixes and their meanings are:

Table 1.3 Commonly used prefixes and their multipliers

Prefix name	Symbol	Multiplier
tera	T	× 10^{12}
giga	G	× 10^{9}
mega	M	× 10^{6}
kilo	k	× 10^{3}
hecto	h	× 10^{2}
deci	d	× 10^{-1}
centi	c	× 10^{-2}
milli	m	× 10^{-3}
micro	μ	× 10^{-6}
nano	n	× 10^{-9}
pico	p	× 10^{-12}

There is really no alternative but to learn these and to practise using them. For example, capacitance values may be given in picofarad (pF) or microfarad (μF). Measurements of length are commonly in kilometre (km), metre (m), centimetre (cm), millimetre (mm) or micrometre (μm). You will need to be able to convert easily from one to another.

> **TIP**
>
> Most prefixes are in multiples of powers of 10^3. If you are not good at converting these yourself, putting a figure, for example, 45 000 000, into your calculator and pressing the 'ENG' button will show it as a power of 10 (but to the nearest multiple of 10^3). Try this and then see the effect of pressing 'shift' 'ENG'.

A Worked examples

a Write the following with a prefix instead of a power of 10:

i **100 000 m**

Step 1: write this as a power of 10.

$100\,000\,\text{m} = 1.0 \times 10^5\,\text{m}$

Step 2: refer to Table 1.3 to see if there is a prefix with this multiplier. If not, rewrite using a close multiplier that is a prefix.

$100\,000\,\text{m} = 100\,\text{km}$ or $0.1\,\text{Mm}$

ii $3.5 \times 10^{-6}\,\text{m}$

Refer to Table 1.3 to see if there is a prefix with this multiplier.

10^{-6} is µ

$3.5 \times 10^{-6}\,\text{m} = 3.5\,\text{µm}$

iii **0.000002 s**

Step 1: write this as a power of 10.

$0.000002\,\text{s} = 2 \times 10^{-6}\,\text{s}$

Step 2: refer to Table 1.3 to see if there is a prefix with this multiplier.

$2 \times 10^{-6}\,\text{s} = 2\,\text{µs}$

iv $5.0 \times 10^{12}\,\text{m}$

Refer to Table 1.3 to see if there is a prefix with this multiplier.

10^{12} is T

$5.0 \times 10^{12}\,\text{m} = 5.0\,\text{Tm}$

v $2.4 \times 10^9\,\text{Hz}$

Refer to Table 1.3 to see if there is a prefix with this multiplier.

10^9 is G

$2.4 \times 10^9\,\text{Hz} = 2.4\,\text{GHz}$

b Write the following in the specified unit with a power of 10 instead of a prefix.

For example: $2.0\,\text{ps (in s)} = 2.0 \times 10^{-12}\,\text{s}$.

i **0.35 cm (in m)**

centi is 10^{-2}

$0.35\,\text{cm} = 0.35 \times 10^{-2}\,\text{m}$ (or $3.5 \times 10^{-3}\,\text{m}$)

ii **180 km (in m)**

kilo is 10^3

$180\,\text{km} = 180 \times 10^3\,\text{m}$ (or $1.80 \times 10^5\,\text{m}$)

iii **900 MHz (in Hz)**

mega is 10^6

$900\,\text{MHz} = 900 \times 10^6\,\text{Hz}$ (or $9.00 \times 10^8\,\text{Hz}$)

iv **600 nm (in m)**

nano is 10^{-9}

$600\,\text{nm} = 600 \times 10^{-9}\,\text{m}$ (or $6.00 \times 10^{-7}\,\text{m}$)

B Guided questions

Copy out the workings and complete the answers on a separate piece of paper.

1 Write the following using a prefix to make the numbers more manageable:

a **1000 m**

Step 1: write this as a power of 10.

$1000\,m = 10^3\,m$

Step 2: refer to Table 1.3 to see if (or remember that) there is a prefix with this multiplier.

$1000\,m = $ _____

b **0.0000001 m**

Step 1: write this as a power of 10.

$0.0000001\,m = 0.1 \times 10^{-6}\,m$

Step 2: refer to Table 1.3 to see if there is a prefix with this multiplier.

$0.0000001\,m = $ _____

Could you use another multiplier instead?

$0.1 \times 10^{-6} = 100 \times 10^{-9}$

c **1 500 000 000 Hz**

Step 1: write this as a power of 10.

$1\,500\,000\,000\,Hz = 1.5 \times 10^9$

Step 2: refer to Table 1.3 to find the prefix for 10^9.

$1\,500\,000\,000\,Hz = $ _____

C Practice questions

2 Write the following without a prefix:
   a 4.5 Mm   b 200 kN   c 800 nm   d 30 MHz

Converting units

In calculations, you will make fewer errors if you always convert quantities to the base SI unit. You will cover these in more detail in your Physics course but the main ones you will encounter at A-level are given in Table 1.4.

Table 1.4 Six of the seven SI base units that you will encounter most frequently at A-level

Base quantity	Name	Symbol
length	metre	m
mass	kilogram	kg
time	second	s
electric current	ampere	A
temperature	kelvin	K
amount of substance	mole	mol

The base unit for mass, the kilogram, is an anomaly in that it already has a prefix associated with it and you need to be aware of this when doing calculations.

When doing a calculation, you will avoid a lot of errors if you get into the habit of converting everything to the base unit as a power of 10. Although this is not always strictly necessary, you will not go wrong if you do this.

For example, when doing a calculation it is easy to change 3.5 cm to m simply by writing 3.5×10^{-2} m or 0.56 mm to m by writing 0.56×10^{-3} m.

Errors often occur when converting areas and volumes. Remember that the area of a square of, for example, 10 cm × 10 cm is 100 cm². A common error is to calculate the area in m² as 100×10^{-2} m², which is wrong. If you convert to the base unit first, you obtain 10×10^{-2} m × 10×10^{-2} m = 100×10^{-4} m², which is correct; there are 1×10^{4} cm² in 1 m².

Similarly for volume, there are 1×10^{6} cm³ in 1 m³ because 1.0 m³ is 100 cm × 100 cm × 100 cm or 10^{2} cm × 10^{2} cm × 10^{2} cm = 10^{6} cm³.

A Worked example

A piece of copper wire is 2.0 cm long and has a diameter of 0.25 mm.

Calculate the volume of copper in m³.

The easiest way, and the way least likely to lead to error, is to convert all measurements to the base unit immediately.

Step 1: identify the base unit and convert all measurements to the base unit.

2.0 cm = 2.0×10^{-2} m

0.25 mm = 0.25×10^{-3} m

Step 2: calculate the cross-sectional area.

The cross-sectional area of the wire is given by area = πr^2, where r is the radius of the wire. However, note that usually a diameter will be given so you can also use area = $\pi \frac{d^2}{4}$ where d is the diameter.

$$\text{cross-sectional area} = \pi \times \frac{(0.25 \times 10^{-3})^2}{4} = 4.9 \times 10^{-8} \text{ m}^2$$

Step 3: calculate the volume.

Volume is given by length × cross-sectional area.

$$\text{volume} = 2.0 \times 10^{-2} \times 4.9 \times 10^{-8} = 9.8 \times 10^{-10} \text{ m}^3$$

TIP

Forgetting to change from diameter to radius is a common error. Practise calculating cross-sectional areas, especially of small objects, so that you are familiar with the process.

B Guided questions

Copy out the workings and complete the answers on a separate piece of paper.

1 **Give the following in the base unit with a power of 10:**

   a **30 cm**

   Step 1: identify the base unit. m

   Step 2: replace the prefix with a power of 10.

   b **0.1 mm**

   Step 1: identify the base unit. m

   Step 2: replace the prefix with a power of 10.

   c **100 g**

   Step 1: identify the base unit. kg

   Step 2: recall that there are 1×10^{-3} kg in one g.

   Step 3: to convert to the base unit for mass, multiply the mass in g by 1×10^{-3}.

2 **Calculate the number of mm³ in 1 m³.**

   There are 1000 or 10^3 mm in 1 m.

   There are $10^3 \times 10^3 \times 10^3$ mm³ in 1 m³.

   As this is a multiplication, add the indices.

3 **The volume of a cylinder is given by $V = \pi r^2 h$, where r is the radius of the cylinder and h is the height of the cylinder.**

   A cylinder has a diameter of 10 cm and a height of 20 cm.

   Calculate the volume in m³.

   Step 1: identify the base unit and convert the diameter and the height to the base unit (remember that centi is 10^{-2}).

   Step 2: calculate the cross-sectional area, writing the radius, r, of the cylinder as $r = \frac{d}{2}$.

   Step 3: calculate the volume.

C Practice questions

4 Calculate the area of the following rectangles in m²:
   a 4.0 m × 5.0 m
   b 6.7 cm × 8.2 cm
   c 4.5 mm × 1.2 cm

5 Calculate the cross-sectional area, in m², of a tube of radius 2.5 cm.

6 Calculate the cross-sectional area, in m², of a wire of diameter 0.49 mm.

7 Calculate the volume in m³ of a wire of diameter 0.35 mm and length 2.5 cm.

8 The resistance, R, of a piece of wire of length, l, and cross-sectional area, A, is given by the formula

$$R = \frac{\rho \times l}{A}$$

where ρ is the resistivity.

Calculate the resistance of a piece of copper wire of length 5.0 m and cross-sectional area $1.9 \times 10^{-7}\,\text{m}^2$.

The resistivity of copper is $1.71 \times 10^{-9}\,\Omega\,\text{m}$.

Symbols

There are a number of mathematical symbols commonly used in A-level Physics. These are shown in Table 1.5.

Table 1.5 Commonly used symbols

Symbol	Meaning	Example
=	Equals	$2\,\text{m s}^{-1} + 8\,\text{m s}^{-1} = 10\,\text{m s}^{-1}$ Shows that two expressions or quantities are equal to each other. Not only must the magnitudes be equal, but the units on each side of the equals sign must be the same. Thus in the familiar equation $s = ut + \frac{1}{2}at^2$, the unit of s (displacement) is metre. Therefore the unit of ut (velocity × time) must also be metre and the unit of $\frac{1}{2}at^2$ (acceleration × time²) must also be metre.
<	Less than	speed of light in water < speed of light in air
<<	Very much less than	speed of sound << speed of light
>	More than	distance between the Earth and the Moon > distance from Earth to a geostationary satellite
>>	Very much more than	Neptune's orbital radius >> Earth's orbital radius
∝	Directly proportional to	$p \propto T$: pressure, p (of an ideal gas), is proportional to the absolute temperature, T. A graph of p against T would be a straight line through the origin. If one quantity doubles, so does the other. Quantities cannot be directly proportional if the graph of one against the other does not pass through the origin. An expression with a proportional sign can be made into an equation by adding a constant, e.g. $p = K \times T$, where K is a constant and must have units Pa K^{-1} in order to make the units on either side of the equation the same. The constant, K, would be the gradient of a graph of p against T.
≈	Approximately equal to	Often used when making order of magnitude calculations or in 'show that' questions. $320\,\text{m s}^{-1} \approx 300\,\text{m s}^{-1}$
Δ	Change in	Used when finding a quantity that depends on a rate of change or the gradient of a graph. For example, $v = \frac{\Delta s}{\Delta t}$ $\left(\text{velocity} = \frac{\text{change of displacement}}{\text{time for change}}\right)$ or $F = \frac{\Delta p}{\Delta t}$ $\left(\text{force} = \frac{\text{change of momentum}}{\text{time for change}}\right)$

> **TIP**
>
> It is a common error to describe quantities as proportional where the relationship is linear but a graph of one against the other does not pass through the origin.

Making sure that the units on each side of an equation are the same can serve as a quick check that an equation could be correct. For example, the equation for the time period of a mass on a spring is $T = 2\pi\sqrt{\frac{m}{k}}$, where k is a spring constant and m is a mass. The term 2π has no units. Therefore the units of $\sqrt{\frac{m}{k}}$ are $\sqrt{\frac{kg}{Nm^{-1}}} = \sqrt{\frac{kg}{kg\,m\,s^{-2}\,m^{-1}}} = \sqrt{s^2} = s$, which is the same unit as the left-hand side of the equation. If you had misremembered the equation as $T = 2\pi\sqrt{\frac{k}{m}}$, a check of the units would quickly show the equation to be wrong since the units would not be the same on both sides of the equation. Do not forget that for this technique to work, the units must all be stated as base units.

B Guided questions

Copy out the workings and complete the answers on a separate piece of paper.

1 **Use units to show that the equation $v^2 = u^2 + 2as$, where v and u are velocities, a is acceleration and s is displacement, could be correct.**

   Step 1: write the base units for the left-hand side of the equation.

   Step 2: write the units for u^2.

   Step 3: work out the units for $2as$.

   Step 4: show that all three sets of units are the same.

2 **Electric field strength, E, is given by $E = \frac{\Delta V}{\Delta x}$, where V is a potential and x is a position. Calculate the electric field strength between two plates which have potentials of +100 V and +200 V and which are positioned at distances of 2.10 cm and 2.20 cm.**

   Step 1: calculate ΔV (remember for change, always calculate final minus initial value).

   Step 2: calculate Δx.

   Step 3: evaluate $\frac{\Delta V}{\Delta x}$.

C Practice questions

3 Calculate the acceleration of a car that increases its velocity from $10\,m\,s^{-1}$ to $13\,m\,s^{-1}$ in 10 s.

4 Charles's law states that $V \propto T$. A sample of gas has a volume of $2.0\,m^3$ at a temperature of 293 K. Calculate the constant of proportionality and rewrite the expression as an equation.

Standard form and significant figures

A number in standard form consists of a figure with one significant figure before the decimal point (i.e. a number between 1 and 10) and a power of 10. For example, a force of 2500 N is written in standard form as 2.5×10^3 N. A stress of 1 000 000 Pa is written as 1.0×10^6 Pa.

To enter either of these into a calculator you would type the number, e.g. 2.5, and then press 'exp' or sometimes '$\times 10^x$' and type the index.

Significant figures

Giving the correct number of significant figures in calculations is important in physics because it signifies the degree of certainty in a value. For example, 2.5×10^3 N means that the value is greater than or equal to 2450 N but less than 2550 N, while 2.50×10^3 N means that the value is greater than or equal to 2495 N but less than 2505 N, thus implying much greater precision.

The use of standard form helps to make clear the number of significant figures and also helps prevent a common error that occurs when re-using a previously calculated figure: if there are a lot of digits, especially zeros, in a number, it is easy to make a mistake and miss one out or reverse some digits. Thus 10 000 000 might be mistaken for 1 000 000, especially if the digits are not correctly spaced.

When deciding on the number of significant figures, leading zeros do not count but trailing zeros do count. For example:
- 0.00345 is given to three significant figures because the leading zeros do not count and there are no zeros after the final 5. It is better written as 3.45×10^{-3}.
- 3450 is given to four significant figures and is better written as 3.450×10^3. Writing 3.45×10^3 is incorrect unless it is known that there is an uncertainty of ±5 in the original number.

The number of figures after the decimal point is not the same as the number of significant figures and should not be confused. Significant figures are all those after any leading zeros in an answer or piece of data. For example:
- 1.24 has three significant figures but two decimal places.
- 123.45 has five significant figures but still has two decimal places.
- 0.0124 has three significant figures but four decimal places.

Worked examples

Put the following into standard form:

a 0.0068

The leading zeros do not count as significant.

$0.0068 \text{ m} = 6.8 \times 10^{-3}$ m

b The speed of light: 300 000 000 m s^{-1}

- This is tricky because the original number is given to nine significant figures so there should be nine significant figures when expressed in standard form.
- In fact, the currently accepted value for the speed of light is 299 792 458 m s^{-1}, which would be 2.9979×10^8 m s^{-1} to five significant figures. However, it is usually given to two or three significant figures because the other data in the calculation is rarely known to more than this precision.
- This shows how using standard form makes it clear how many significant figures are being given.

300 000 000 m s^{-1} = 3.00×10^8 m s^{-1} to three significant figures

c 0.250 mm

The trailing 0 at the end is significant and should be included.

$0.250 \text{ mm} = 2.50 \times 10^{-1}$ mm or 2.50×10^{-4} m

When doing calculations, you should always give your answer to no more significant figures than the least number in the data given. For example, an area of land 100 m × 20 m has an area 2.0×10^3 m². Writing 2000 m² would be incorrect as this implies four significant figures whereas the minimum number of significant figures in the data is two. However, it is acceptable to write '2000 m² (2 s.f.)' to show you are claiming a precision of two significant figures.

> **REMEMBER**
>
> You will lose marks if you do not give your answers to the correct number of significant figures.

If data is given to three significant figures, your calculation answer should also be to three significant figures and you should not write, for example, 3.56 (2 d.p.) where the (2 d.p.) stands for two decimal places. It is the number of significant figures that is important, not the number of decimal places. 3.56 cm could be written 0.0356 m. Both have three significant figures but, in the first case, two decimal places and in the second case, four decimal places.

When doing calculations where numbers are taken from one stage to the next, you should use all the figures from each stage of a calculation at the next stage and only reduce your answer to the appropriate number of significant figures at the end.

When rounding your answer to fewer significant figures than you have on your calculator, remember that 5 or above should be rounded up and below 5 should be rounded down. If you want to round to two significant figures, you need to look at the next significant figure (the third one) and decide if it should be rounded up or not. For example:

- 1.2445 rounded to two significant figures is 1.2 because the third significant figure (the first 4) is less than 5 and so is rounded down.
- 1.2445 rounded to four significant figures is 1.245 because the fifth significant figure (5) is rounded up.
- 1.2545 rounded to two significant figures is 1.3 because the third significant figure (the first 5) is equal to 5 and so is rounded up.

A Worked examples

a A current of 1.45 A flows through a resistor of 6.8 Ω. Calculate the potential difference across the resistor.

To calculate the potential difference you need to use the formula $V = I \times R$, where V is the voltage, I is the current and R is the resistance.

Step 1: substitute the values in the formula.

$V = I \times R$

$= 1.45 \times 6.8$

$= 9.86$ V

Step 2: as the minimum number of significant figures in the original data is two, give the answer to two significant figures.

$V = 9.9$ V

b **The radius of Venus is 6052 km. Give this radius to two significant figures.**

Step 1: the third significant figure is 5. This is rounded up.

Step 2: write the number to two significant figures.

6.05×10^3 km to two significant figures is 6.1×10^3 km.

c **Show that the area of a capacitor plate with dimensions 2.54 cm × 2.0 m is about 0.05 m².**

Step 1: convert 2.54 cm to base units.

2.54×10^{-2} m

Step 2: find the area by multiplying.

$2.54 \times 10^{-2} \times 2.0 = 0.0508$ m²

Step 3: You are asked to 'show' and are given a value to one significant figure. Give your answer to at least one more significant figure.

area of the capacitor plate = 0.051 m²

TIP

In 'show that' questions, always give your answer to one more significant figure than the value you are given.

B Guided questions

Copy out the workings and complete the answers on a separate piece of paper.

1 **An image sensor chip in a camera has dimensions 5.76 mm × 4.29 mm. Calculate the area of the sensor in mm² to the correct number of significant figures.**

Step 1: find the area by multiplying the two dimensions.

area of sensor = $5.76 \times 4.29 = 24.7104$ mm²

Step 2: as the minimum number of significant figures in the original data is three, give the answer to three significant figures.

area of sensor = _____ mm²

2 **Calculate the number of seconds in a year, assuming that a year is 365 days. Give your answer in standard form and to the correct number of significant figures.**

- The number of hours in a day is 24 (to many significant figures).
- The number of minutes in an hour and the number of seconds in a minute are also defined.
- The limiting data is 365 days in a year since this is given to just three significant figures.

Step 1: calculate the number of seconds in one day.

$$\text{number of seconds in a day} = \text{number of seconds in a minute} \times \text{number of minutes in an hour} \times \text{number of hours in a day}$$

Step 2: multiply your answer to Step 1 by the number of days in a year (365).

number of seconds in a year = _____ × 365

= _____ s

Step 3: present your answer in standard form.

C Practice questions

3 An oxygen atom has a diameter of 120 pm.

Write this diameter in standard form.

4 The distance from the Earth to the Moon is 384 400 km.

Write this distance in standard form, in metres, to three significant figures.

5 Put the following into standard form with the correct number of significant figures:
   a 0.000783
   b 1470
   c 0.000000056

6 It takes 4.24 years for light to reach Earth from the star Proxima Centauri.

Light travels at $3.00 \times 10^8 \, \text{m s}^{-1}$. There are $3.15 \times 10^7 \, \text{s}$ in a year.

Calculate the distance to Proxima Centauri in metres. Give your answer in standard form to the correct number of significant figures.

Calculations using standard form

If you want to multiply two numbers in standard form, for example $(4.0 \times 10^3) \times (5.0 \times 10^4)$, you can key in both numbers to a calculator and get 2.0×10^8. However, it is just as easy to multiply the first part of each number and add the indices of the second part. This becomes:

$$4.0 \times 5.0 \times (10^{(3+4)}) = 20 \times 10^7$$

This is not in standard form so needs to be rewritten as 2.0×10^8.

Dividing using standard form is just as easy but here you divide the first part of each number and subtract the indices of the second part. For example, $\frac{5.0 \times 10^4}{4.0 \times 10^3}$ becomes:

$$\frac{5}{4} \times (10^{(4-3)}) = 1.25 \times 10^1 \text{ or } 12.5$$

Since the original numbers are given to only two significant figures, the answer should also be given to two significant figures and would be 1.3×10^1.

A more complex calculation is also made easier in this way, for example:

$$\frac{5.0 \times 10^4 \times 4.0 \times 10^3}{2.0 \times 10^{-3}} = \frac{5.0 \times 4.0}{2.0} \times 10^{(4+3-(-3))} = 10 \times 10^{10}$$

In standard form this would be written 1.0×10^{11}.

A Worked example

The Young modulus of an elastic material is given by the formula $E = \frac{F \times l}{A \times e}$.

F is the force that causes extension, e, in a wire of length l and cross-sectional area A.

A wire of length 2.0 m extends by 3.0 mm when a force of 8.0 N is applied. The cross-sectional area of the wire is $4.9 \times 10^{-8}\,\text{m}^2$.

Calculate the Young modulus.

Step 1: convert to base units where necessary.

$3.0\,\text{mm} = 3.0 \times 10^{-3}\,\text{m}$

Step 2: substitute the given values into the formula.

$E = \frac{F \times l}{A \times e} = \frac{8.0 \times 2.0}{4.9 \times 10^{-8} \times 3.0 \times 10^{-3}} = 1.088\ldots \times 10^{11}\,\text{N m}^{-2}$

Step 3: check that the answer is roughly correct and is written in standard form.

A quick check of the calculation (with all values rounded to only one significant figure) shows that the value is approximately $\frac{20}{15} \times \frac{1}{10^{-11}}$, which is approximately 1×10^{11} — close to the calculated answer.

With a bit of practice you will be able to estimate an answer so you will know if it is approximately correct. This provides a quick check in case you key in the figures incorrectly to your calculator.

Step 4: check the number of significant figures. The data in the question is to two significant figures so your answer should be to two significant figures only, which is 1.1×10^{11}.

Step 5: add the correct unit.

N m^{-2} is the same as Pa (pascal) in SI units, so the Young modulus is

$1.1 \times 10^{11}\,\text{Pa}$

TIP

Get into the habit of checking approximate answers to ensure the answer you write down from your calculator is sensible.

B Guided questions

Copy out the workings and complete the answers on a separate piece of paper.

1. The electric field strength, E, between two plates a distance d apart and that have a potential difference of V across them is given by $E = \frac{V}{d}$.

   A pair of plates are 2.5 mm apart and have a potential difference of 27 V across them.

   Calculate the electric field strength between the plates, giving your answer in standard form to the correct number of significant figures.

   Step 1: convert to base units where necessary.

   $2.5\,\text{mm} = 2.5 \times 10^{-3}\,\text{m}$

Step 2: substitute the given values into the formula.

$$E = \frac{27}{2.5 \times 10^{-3}}$$

Step 3: check that the answer is roughly correct and is written in standard form.

Your calculator answer will be 10 800. This is not in standard form so convert to standard form with one digit before the decimal place.

A rough estimate of the answer gives _____.

Step 4: check the number of significant figures. The data in the question is to two significant figures so your answer should be to two significant figures only.

Step 5: add the correct unit. _____ V m^{-1}

2 **The force, F, on a wire of length l carrying a current I at right angles to a magnetic field of flux density B is given by $F = BIl$.**

A wire of length 5.0 cm carries a current of 4.5 A in a field of flux density 3.4 mT.

Calculate the force on the wire, giving your answer in standard form to the correct number of significant figures.

Step 1: convert to base units where necessary.

   5.0 cm = 5.0×10^{-2} m

Replace the prefix in the value of the flux density to give 3.4×10^{-3} T.

Step 2: substitute the given values into the formula.

   $F = 3.4 \times 10^{-3} \times 4.5 \times 5.0 \times 10^{-2}$

Step 3: check that the answer is roughly correct and is written in standard form.

Step 4: check the number of significant figures. The data is given to two significant figures so you need to round your answer to two significant figures.

Step 5: add the correct unit for force.

3 **The average velocity, v, of a body is given by $v = \frac{\Delta s}{\Delta t}$, where Δs is the change in displacement and Δt is the time over which the change happens.**

A boat travels a distance of 36.5 km in 45.0 min in a straight line.

Calculate its average velocity in m s^{-1}, giving your answer in standard form to the correct number of significant figures.

Step 1: convert to base units where necessary.

   36.5 km = _____

   45 min = _____

Step 2: substitute the given values into the formula.

   $v = \frac{\Delta s}{\Delta t}$

Step 3: check that the answer is roughly correct and is written in standard form.

Step 4: check the number of significant figures. The data is given to three significant figures, so round your answer to three significant figures.

Step 5: add the correct unit for velocity.

> **TIP**
>
> When doing calculations that involve separate calculations on your calculator, always use brackets around the separate calculations to avoid significant errors.

C Practice questions

4 Calculate the following and give the answer in standard form to two significant figures:

a $\dfrac{4.5 \times 6.3}{5.4}$

b $9.1 \times 10^5 \times 2.3 \times 10^{-2}$

c $\dfrac{8.9 \times 10^{12}}{3.0 \times 10^8}$

d $1.6 \times 10^{-19} \times 56$

e $\dfrac{23}{56 \times 26}$

f $\dfrac{9.0 \times 10^9 \times (1.6 \times 10^{-19})^2}{1.0 \times 10^{-10}}$

5 The pressure, p (in Pa), of a volume, V, of n moles of an ideal gas at an absolute temperature, T, is given by $P = \dfrac{nRT}{V}$, where R is a constant of value $8.31\,\text{J}\,\text{K}^{-1}\,\text{mol}^{-1}$.

A container of ideal gas contains 136 moles and has a volume of $3.0\,\text{m}^3$. The temperature of the gas is $294\,\text{K}$.

Calculate the pressure of the gas in the container. Give your answer in standard form to the correct number of significant figures.

6 The tensile stress, σ, in a wire of cross-sectional area, A, when it experiences a tensile force, F, is given by $\sigma = \dfrac{F}{A}$.

A wire of diameter $2.4\,\text{mm}$ is stretched with a force of $7.84\,\text{N}$.

Calculate the tensile stress in the wire. Give your answer in standard form to the correct number of significant figures.

Orders of magnitude

$1\,000\,000\,\text{Pa}$ can be abbreviated to $1.0\,\text{MPa}$ or to $10^6\,\text{Pa}$. This last abbreviation is usually called an order of magnitude. Orders of magnitude are used frequently to make quick estimates. Making orders of magnitude estimates is not easy but it is worthwhile trying to practise the skill by making simple estimates of sizes and values in the world around you. Getting a 'feel' for the size of things and for numbers is a useful skill to acquire and, by practising, you will get better at it. In some questions you will be asked to give an order of magnitude for a quantity, so it is a good idea to try to become familiar with the magnitude of quantities you are working with.

Orders of magnitude are values given to the nearest power of 10. For example, the weight of a person would be $1000\,\text{N}$ or $10^3\,\text{N}$ as an order of magnitude because they are clearly not $1\,\text{N}$ (approximately the weight of an apple), $10\,\text{N}$ (approximately the weight of a bag of sugar), 100 N (approximately the weight of hand baggage allowed on most aircraft) or 10 000 N (approximately the weight of a small car). This thinking process shows the way you might go about making an order of magnitude estimate, which is by thinking of something that is bigger, the same size or smaller, and going for an order of magnitude in between something bigger and something smaller.

For example, to give an order of magnitude estimate for the height of a large tree, think of something you might know. A block of flats that is the same height as a large tree might have four storeys; a room is taller than you, about 2.5 m high, so a sensible order of magnitude estimate for the height of a tree would be 10 m.

Now think of an order of magnitude estimate for the thickness of a piece of paper. It must be thinner than a coin, which is about a millimetre, and it would take several pieces of paper to have the same thickness as a coin. We can therefore give an order of magnitude for the thickness of paper as 0.1 mm or 10^{-4} m.

Interestingly, the current record claimed for the number of times it is possible to fold a long, narrow piece of paper in half is 12 times. The final thickness of paper 0.1 mm thick when folded 12 times is about 40 cm! For an ordinary piece of A4 paper, the limit is seven or eight folds. You can find out more about this by searching online for 'maximum paper folding'.

Another example of an order of magnitude estimate is to estimate the time taken to drive from Edinburgh to London in the UK. You will need to know, or look up, the distance from Edinburgh to London by road, which is about 400 miles or 700 km. To the nearest order of magnitude, this is 1000 km or 10^6 m.

In the UK, the speed limit for dual carriageway roads is 70 mph, which is about 100 km h^{-1} since 1 km is about 0.6 of a mile. If a car could travel at an average speed of about 100 km h^{-1} for the whole journey, the journey from Edinburgh to London would take about $\frac{1000 \text{ km}}{100 \text{ km h}^{-1}}$ = 10 hours. This type of order of magnitude estimate can be surprisingly good: a mapping program gives the continuous journey time by car from Edinburgh to London as about 7.5 h.

The process of making order of magnitude estimates might sound complicated but it can be useful for checking that answers to problems are sensible or for making rough estimates when planning experiments. It is the sort of thing you can practise when waiting in the lunch queue or at the bus stop. It is a good idea to think quantitatively and try to notice and remember some values for familiar objects such as the mass of a bag of sugar, the density of air, the typical mass of a person etc.

Worked examples

a **Give an order of magnitude value for the mass of a textbook in kg.**

Step 1: think of a known thing that is bigger, the same size or smaller.

Decide which is best between 0.1 kg (mass of an apple), 1 kg (mass of a bag of sugar) and 10 kg (a full load of wet washing in a washing machine).

Step 2: choose the most appropriate order of magnitude.

As an order of magnitude estimate, the mass of a textbook in kg is closest to 1 kg.

b **Make an order of magnitude estimate of the mass of air in a physics lab.**

Of course this depends on the size of the lab. However, you might go about this estimate in the following way:

Step 1: break down the order of magnitude estimate into stages.

- Find an order of magnitude estimate for the volume of the lab. A lab designed for 30 students might have a floor area of about $10\,m \times 4\,m$ and a ceiling height of $2.5\,m$, giving a volume of $10 \times 4 \times 2.5 = 100\,m^3$.
- Even if you do not have an idea of the density of air, you probably know that 1 mole of any gas at standard temperature and pressure (stp) occupies $22.4\,dm^3$.
- 1 mole of nitrogen has a mass of about 28 g, therefore the density, ρ, of nitrogen ≈ density of air and is $\dfrac{28 \times 10^{-3}\,kg}{22.4 \times 10^{-3}\,m^3} = 1.25\,kg\,m^{-3} \approx 1\,kg\,m^{-3}$

Step 2: put together the various order of magnitude estimates.

mass of air = volume × density = $100\,m^3 \times 1\,kg\,m^{-3} = 100\,kg$

B Guided questions

Copy out the workings and complete the answers on a separate piece of paper.

1 **Give an order of magnitude value for the mass of a car in kg.**

Step 1: think of a known thing that is bigger, the same size or smaller.

Decide if a car is nearest to 100 kg (about three standard paving slabs), 1000 kg/ 1 tonne ($1\,m^3$ of water) or 10 000 kg (a bit more than an African elephant).

Step 2: choose the most appropriate order of magnitude.

Order of magnitude of the mass of a car is _____ kg.

2 **Give an order of magnitude value for the diameter of an atom.**

Step 1: break down the order of magnitude estimate into stages.

- The diameter of an oxygen atom will not be significantly different from the diameter of a water molecule. The molar mass of water is 18 g. This means that in 1000 kg (which is $1\,m^3$) there are $\dfrac{1000}{18 \times 10^{-3}} = 56 \times 10^3$ moles ≈ 100 000 moles.
- The Avogadro number is the number of molecules in a mole. It is $6.02 \times 10^{23} \approx 10 \times 10^{23}$.
- $1\,m^3$ of water contains approximately $100\,000 \times 10 \times 10^{23} = 10^{29}$ molecules.
- The volume occupied by one molecule is therefore $\dfrac{1}{10^{29}} = 10^{-29}\,m^3$.

Step 2: put together the various order of magnitude estimates. Assume that each molecule occupies a cube into which it just fits. One side would be $\sqrt[3]{10^{-29}} \approx 10^{-10}\,m$ and this is the approximate diameter of a water molecule, which is not significantly different to the diameter of an oxygen atom.

Order of magnitude of the diameter of an atom is _____ m.

C Practice questions

3 Give an order of magnitude value for the wavelength of light.

4 Make an order of magnitude estimate of the kinetic energy of a car travelling at about 30 mph.

The formula for kinetic energy is k.e = $\tfrac{1}{2}mv^2$.

2 Fractions, ratios and percentages

In A-level Physics you will frequently have to use fractions, ratios and percentages. For example, the refractive index of a material is the ratio of two speeds and the tensile strain in a stretched wire is the ratio of extension to original length and is expressed both as a fraction and as a percentage.

Fractions

The most commonly used calculators will give answers as a fraction, e.g. $\frac{5}{24}$, although you can set your calculator not to do so. To convert from a fraction to a decimal, you need to divide the numerator (top number) by the denominator (bottom number). In this case, the answer is 0.208 to three significant figures. Your calculator will do this calculation for you with the press of a single button. On many calculators, this button is labelled $S \Leftrightarrow D$.

When multiplying fractions, you multiply the numerators together and then the denominators together, for example, $\frac{2}{3} \times \frac{5}{6} = \frac{(2 \times 5)}{(3 \times 6)} = \frac{10}{18}$. In this case we can then simplify the fraction by dividing both top and bottom by 2 to give $\frac{5}{9}$.

When dividing fractions, for example, $\frac{2/3}{5/6}$, notice that dividing by $\frac{5}{6}$ is the same as multiplying by $\frac{1}{5/6} = \frac{6}{5}$. Therefore $\frac{2/3}{5/6} = \frac{2}{3} \times \frac{6}{5} = \frac{12}{15}$. In this case we can divide both top and bottom by 3 to give $\frac{4}{5}$.

When adding or subtracting fractions, the denominator must be the same before the fractions can be added or subtracted. For example, $\frac{2}{3} + \frac{5}{6}$ cannot be added directly because the denominators are different. However, if we multiply the top and bottom of $\frac{2}{3}$ by 2, we get the equivalent fraction $\frac{4}{6}$ and the addition becomes $\frac{4}{6} + \frac{5}{6} = \frac{(4+5)}{6} = \frac{9}{6}$ which, if we divide both top and bottom by 3, becomes $\frac{3}{2}$. Similarly $\frac{2}{3} - \frac{5}{6}$ becomes $\frac{4}{6} - \frac{5}{6} = \frac{(4-5)}{6} = -\frac{1}{6}$.

Sometimes when adding or subtracting fractions we have to multiply the top and bottom of each fraction by a different number in order to obtain what is called the lowest common denominator. For example, to calculate $\frac{3}{5} + \frac{2}{3}$ we need to multiply the top and bottom of $\frac{3}{5}$ by 3 to give $\frac{9}{15}$ and the top and bottom of $\frac{2}{3}$ by 5 to give $\frac{10}{15}$. The addition then becomes $\frac{9}{15} + \frac{10}{15} = \frac{(9+10)}{15} = \frac{19}{15}$. Exactly the same rules apply when adding, subtracting or multiplying algebraic fractions.

A Worked examples

a   Multiply the following fractions:

  i $\frac{4}{7} \times \frac{5}{3}$

  Step 1: multiply the numerators. $4 \times 5 = 20$

  Step 2: multiply the denominators. $7 \times 3 = 21$

  Step 3: write as a single fraction. $\frac{20}{21}$

  ii $\frac{x}{y} \times \frac{a}{b}$

  Step 1: multiply the numerators. $x \times a = xa$

  Step 2: multiply the denominators. $y \times b = yb$

  Step 3: write as a single fraction. $\frac{xa}{yb}$

b   Divide the following fractions:

  i $\frac{5/7}{4/3}$

  Step 1: turn the division into a multiplication. $\frac{5/7}{4/3} = \frac{5}{7} \times \frac{3}{4}$

  Step 2: multiply the numerators and multiply the denominators.

  $$\frac{5}{7} \times \frac{3}{4} = \frac{5 \times 3}{7 \times 4}$$

  Step 3: write as a single fraction (simplify where possible).

  $$\frac{5 \times 3}{7 \times 4} = \frac{15}{28}$$

  ii $\frac{(a+b)}{x} \div \frac{c}{y}$

  Step 1: turn the division into a multiplication.

  $$\frac{(a+b)}{x} \div \frac{c}{y} = \frac{(a+b)}{x} \times \frac{y}{c}$$

  Step 2: multiply the numerators and multiply the denominators.

  $$\frac{(a+b)y}{xc}$$

c   Add the following:

  i $\frac{5}{7} + \frac{2}{3}$

  Step 1: find a common denominator. 21 is the smallest number that both 7 and 3 divide into.

  Step 2: multiply the top and bottom of each fraction to write it using the common denominator.

  Multiply the top and bottom of $\frac{5}{7}$ by 3 to give $\frac{15}{21}$.

  Multiply the top and bottom of $\frac{2}{3}$ by 7 to give $\frac{14}{21}$.

  Step 3: add the fractions.

  $$\frac{15}{21} + \frac{14}{21} = \frac{(15+14)}{21} = \frac{29}{21}$$

ii $\quad \dfrac{1}{C_1} + \dfrac{1}{C_2}$

Step 1: find a common denominator. This will be $C_1 C_2$.

Step 2: multiply the top and bottom of each fraction to write it using the common denominator.

Multiply the top and bottom of $\dfrac{1}{C_1}$ by C_2 to give $\dfrac{C_2}{C_1 C_2}$.

Multiply the top and bottom of $\dfrac{1}{C_2}$ by C_1 to give $\dfrac{C_1}{C_1 C_2}$.

Step 3: add the fractions.

$$\dfrac{C_2}{C_1 C_2} + \dfrac{C_1}{C_1 C_2} = \dfrac{(C_1 + C_2)}{C_1 C_2}$$

B Guided questions

Copy out the workings and complete the answers on a separate piece of paper.

1 **Add the following fractions:**

  a $\dfrac{3}{5} + \dfrac{11}{15}$

  Step 1: find a common denominator.

  Step 2: multiply the top and bottom of each fraction to write it using the common denominator.

  Multiply the top and bottom of $\dfrac{3}{5}$ by 3 to give $\dfrac{9}{15}$.

  Step 3: add the fractions. $\dfrac{9}{15} + \dfrac{11}{15} =$ _____

  Step 4: simplify the final fraction if possible.

  b $\dfrac{3}{8} + \dfrac{5}{9}$

  Step 1: find a common denominator.

  Step 2: multiply the top and bottom of each fraction to write it using the common denominator.

  Multiply the top and bottom of $\dfrac{3}{8}$ by 9 to give $\dfrac{27}{72}$.

  Multiply the top and bottom of $\dfrac{5}{9}$ by 8 to give $\dfrac{40}{72}$.

  Step 3: add the fractions. $\dfrac{27}{72} + \dfrac{40}{72} =$ _____

  c $\dfrac{1}{v^2} + \dfrac{1}{c^2}$

  Step 1: find a common denominator.

  Step 2: multiply the top and bottom of each fraction to write it using the common denominator.

  Multiply the top and bottom of $\dfrac{1}{v^2}$ by c^2 to give $\dfrac{c^2}{v^2 c^2}$.

  Multiply the top and bottom of $\dfrac{1}{c^2}$ by v^2 to give $\dfrac{v^2}{v^2 c^2}$.

  Step 3: add the fractions.

2 Fractions, ratios and percentages

2 **Calculate the following:**

a $\dfrac{3}{5} \times \dfrac{2}{3}$

Step 1: multiply the numerators.

Step 2: multiply the denominators.

Step 3: write as a single fraction.

b $\dfrac{6}{7} \div \dfrac{3}{4}$

Step 1: turn the division into a multiplication.

Dividing by $\dfrac{3}{4}$ is the same as multiplying by $\dfrac{4}{3}$.

$\dfrac{6}{7} \times \dfrac{4}{3}$

Step 2: multiply the numerators and multiply the denominators.

Step 3: write as a single fraction.

c $\dfrac{mv^2/r^2}{2\pi e/T}$

Step 1: turn the division into a multiplication.

Dividing by $\dfrac{2\pi e}{T}$ is the same as multiplying by $\dfrac{T}{2\pi e}$.

$\dfrac{mv^2}{r^2} \times \dfrac{T}{2\pi e}$

Step 2: multiply the numerators and multiply the denominators.

Step 3: write as a single fraction.

Practice questions

3 Evaluate the following:

a $\dfrac{3}{4} \times \dfrac{3}{8}$

b $\dfrac{1}{r} \times \dfrac{GMm}{r}$

c $\dfrac{F}{A} \div \dfrac{e}{l}$

d $\dfrac{5}{6} + \dfrac{3}{7}$

e $\dfrac{1}{R_1} + \dfrac{1}{R_2}$

4 The radius, r, of a piece of wire is given as $\dfrac{3}{8}$ mm.

Calculate the value of r^2 as a fraction.

5 The average power in a resistor is $P = \dfrac{V_0}{\sqrt{2}} \times \dfrac{I_0}{\sqrt{2}}$

Write $\dfrac{V_0}{\sqrt{2}} \times \dfrac{I_0}{\sqrt{2}}$ as a single fraction.

Ratios

A ratio is the number of times a quantity is bigger or smaller than another, reference quantity. For example, a step-up transformer may have a turns ratio $N_s : N_p$ of $3:1$, which means that there are three times as many turns on the secondary coil, N_s, as there are on the primary coil, N_p. This would be stated as 'the turns ratio in the step-up transformer is 3'. The turns ratio of a step-down transformer $N_s : N_p$ might be $1:3$, in which case there are three times as many turns on the primary coil as on the secondary coil. It could also be given as a fraction: $\frac{1}{3}$. For a ratio to be valid, the quantities being compared must be of the same unit. Thus a ratio, even when expressed as a fraction or decimal, does not have a unit.

A Worked examples

The ratio of voltages in an ideal transformer is the same as the ratio of the number of turns on the primary and secondary coil.

A transformer has 120 turns on the primary coil and 25 turns on the secondary coil. The voltage in the primary coil is 24 V.

a **Calculate the ratio $N_s : N_p$.**

Divide the number of turns in the secondary coil by the number of turns in the primary coil.

The ratio is $\frac{25}{120} = \frac{5}{24}$, therefore the ratio $N_s : N_p$ is $5:24$ (or it could be given as $\frac{5}{24} = 0.21$).

b **Calculate the voltage in the secondary coil.**

The ratio of voltages, $V_s : V_p$, is the same as the ratio $N_s : N_p$, $5:24$, therefore $\frac{V_s}{V_p} = \frac{V_s}{24} = \frac{5}{24}$ and so, multiplying both sides by 24, we get $V_s = \frac{5 \times 24}{24} = 5$ V.

B Guided questions

Copy out the workings and complete the answers on a separate piece of paper.

1. The speed of light in air is 3.0×10^8 m s^{-1}. The speed of light in a certain type of glass is 2.0×10^8 m s^{-1}.

    Calculate the ratio of the speed of light in air to the speed of light in glass.

    Divide the speed of light in air by the speed of light in glass. Note that there is no unit.

    $$\text{Ratio} = \frac{\text{speed of light in air}}{\text{speed of light in glass}} = \underline{\qquad}$$

2. A piece of wire 1.5 m long is loaded with a mass causing it to stretch by 1.0 mm.

    Calculate the ratio of the extension to the original length.

    Step 1: ensure that the two quantities are in the same units.

    $1.0\,\text{mm} = 1.0 \times 10^{-3}$ m

    Step 2: divide the extension by the original length. (This is generally given as a decimal.)

    $$\text{Ratio} = \frac{\text{extension}}{\text{original length}} = \underline{\qquad}$$

C Practice questions

3. Calculate the ratio of turns, $N_s : N_p$, on a step-up transformer that has 900 turns on the primary coil and 22 500 turns on the secondary coil.

4. The speed of light in water is $2.3 \times 10^8 \, \text{m s}^{-1}$. The speed of light in a certain type of glass is $2.0 \times 10^8 \, \text{m s}^{-1}$.

   Calculate the ratio of the speed of light in water to the speed of light in glass.

Percentages

At A-level, you will most often be required to calculate percentage changes, for example, the percentage increase in length or the percentage uncertainty in a measurement. Efficiencies are also usually given as percentages.

A percentage is really just a fraction: $x\%$ of a quantity is x hundredths of that quantity. The percentage that one quantity is of another is found by calculating

$$\frac{\text{new value}}{\text{original value}} \times 100\%$$

This is particularly useful for efficiencies of an energy transfer process:

$$\text{efficiency} = \frac{\text{useful energy output}}{\text{total energy input}} \times 100\%$$

Calculating a percentage change is straightforward if you remember two simple rules. The first rule is that:

$$\% \text{ change} = \frac{\text{change in value}}{\text{original value}} \times 100\%$$

The second rule is that a change in value is always the final value minus the initial value, even if the final value is smaller than the initial value.

A Worked examples

a **A lift with passengers of total mass, m, of 800 kg is lifted through a height, h, of 10 m by an electric motor.**

   The mechanical work done on the lift is given by work done = mgh, where g is the gravitational field strength of $9.8 \, \text{N kg}^{-1}$.

   The electrical work done on the motor to raise the lift is $1.0 \times 10^5 \, \text{J}$.

   Calculate the efficiency of the lift system.

   Use the formula for calculating efficiency.

   The useful work (output) is $mgh = 800 \times 9.8 \times 10 \, \text{J} = 78\,400 \, \text{J}$

   $$\text{efficiency} = \frac{\text{output (useful) work}}{\text{input}} \times 100\% = \frac{78\,400}{1.0 \times 10^5} \times 100\% = 78\%$$

Full worked solutions at www.hoddereducation.co.uk/essentialmathsanswers

b A car increases its speed from 10 m s⁻¹ to 12 m s⁻¹.

Calculate the percentage change in speed.

Step 1: calculate the change in value as (final value − initial value).

change in speed = $(12 - 10)$ m s⁻¹

Step 2: use the formula for percentage change.

$$\% \text{ change} = \frac{\text{change in value}}{\text{original value}} \times 100\% = \frac{12-10}{10} \times 100\% = 20\%$$

Sometimes you will be asked to calculate a new value given a percentage change. In order to do this, you need to remember that a percentage is a fraction. For example, an uncertainty of 8% in a length of 1.5 m is calculated as $\frac{8}{100} \times 1.5 = 0.12$ m. With practice you will be able to do the conversion of a percentage to a fraction or decimal, i.e. 8% → 0.08, in your head as you carry out a calculation. Occasionally you may be given a new value and told that this represents a certain percentage change from the previous value, which you will then be asked to calculate.

Worked example

A loaded wire is measured to be 2.015 m long. When the load was added, its length increased by 2.5%.

Calculate the original length of the wire.

Step 1: take the original length to be x.

Step 2: calculate the change in value as (final value − initial value).

change in length = $(2.015 - x)$

Step 3: use the formula for calculating a percentage change.

$$\% \text{ change} = \frac{\text{change in value}}{\text{original value}} \times 100\%$$

$$\frac{(2.015 - x)}{x} \times 100\% = 2.5\%$$

Step 4: write the percentage as a fraction and solve the equation.

2.5% as a fraction is $\frac{2.5}{100} = 0.025$.

Therefore $\frac{(2.015 - x)}{x} = 0.025$.

- Multiply both sides by x. (See the section on re-arranging equations in Unit 4.)

  $2.015 - x = 0.025x$

- Add x to both sides. $2.015 = x + 0.025x = 1.025x$

- Divide both sides by 1.025 to give $x = \frac{2.015}{1.025} = 1.96585 \approx 1.966$ m.

- Since the percentage change was given to only two significant figures, we should write the answer to two significant figures.

  The original length was 2.0 m.

You can see from this example that simply dividing the final value by (1 + fractional change) gives the original value. Remember this useful shortcut.

B Guided questions

Copy out the workings and complete the answers on a separate piece of paper.

1. An electric motor does 150 J of mechanical work when it has an input of electrical work of 200 J.

   Calculate the efficiency of the motor.

   Use the formula for calculating efficiency. The useful energy output is 150 J.

   $$\text{efficiency} = \frac{\text{useful energy output}}{\text{total energy input}} \times 100\%$$

   % efficiency = _____

2. A student measures the voltage across a cell as 1.23 V. The cell is being recharged and a short time later the voltage across it is measured to be 1.26 V.

   Calculate the percentage change in voltage.

   Step 1: calculate the change in voltage using (final voltage − initial voltage).

   Step 2: use the percentage change formula.

   $$\% \text{ change} = \frac{\text{change in voltage}}{\text{original voltage}} \times 100\% = \underline{\qquad}$$

3. An ammeter is known to read 5% higher than the actual value. A student measures the current in a circuit as 0.83 A.

   Calculate the actual value of the current.

   Step 1: write the percentage change as a fraction.

   $$5\% = \frac{5}{100} = 0.05$$

   Step 2: remember the shortcut of dividing the final value by (1 + fractional change) to get the original value.

   actual value = _____

C Practice questions

4. The Earth is a slightly 'squashed' sphere. The mean polar radius is 6356.8 km and the mean equatorial radius is 6378.1 km.

   Calculate the percentage difference between the polar radius and the equatorial radius.

5. A student measures the current in a circuit to be 2.35 A with one ammeter. Using a different, more reliable, meter the current reading is 2.30 A.

   Calculate the percentage error in the student's measurement using the first meter.

6. A space rocket fires its thruster and increases its speed by 16%. The final speed is 1500 m s^{-1}.

   Calculate the original speed of the space rocket.

7. A wire of length 2.50 m is stretched by 0.30%.

   Calculate the new length of the wire.

8. A steam engine burns 2.0 tonnes of coal and does 2.1×10^{10} J of useful mechanical work. When burnt in air completely, 1 tonne of coal is able to do 3.0×10^{10} J of work.

   Calculate the percentage efficiency of the steam engine.

3 Averages and probability

Averages

An average, or mean, is useful when taking experimental results. You may take several values for the same measurement and find the mean. To find the mean, you add all the values together and divide by the number of values you have added.

 Worked example

A student measures the time for ten swings of a pendulum, ten times. The values obtained are:

8.0 s, 8.2 s, 8.1 s, 7.8 s, 7.8 s, 8.1 s, 8.2 s, 8.0 s, 7.9 s, 8.1 s.

Calculate the mean time for ten swings and the mean time for one swing.

Step 1: add the values together.

$8.0 + 8.2 + 8.1 + 7.8 + 7.8 + 8.1 + 8.2 + 8.0 + 7.9 + 8.1 = 80.2$

Step 2: divide by the number of values.

mean time for ten swings =

$$\frac{8.0 + 8.2 + 8.1 + 7.8 + 7.8 + 8.1 + 8.2 + 8.0 + 7.9 + 8.1}{10} = \frac{80.2}{10} = 8.02 \text{ s}$$

Therefore the mean time for one swing is $\frac{8.02}{10} = 0.802$ s.

When you are finding the mean, you might expect the final result to be between two of the values given, since the values that are obtained are not all exactly the same and there is no guarantee that the sum of the values will divide neatly by the number of values. It is therefore normal to give a mean value to one more significant figure than the data from which you are finding the mean.

The other two terms you should know in the context of finding a mean are the mode and the median.

The mode is the most frequently occurring result. In the set of numbers in the worked example above, the mode would be 8.1 s.

The median is the middle result. You spread the results in order of value, including numbers that are the same. Using the set of results from the worked example and putting them in order we get: 7.8 s, 7.8 s, 7.9 s 8.0 s, 8.0 s, 8.1 s, 8.1 s, 8.1 s, 8.2 s, 8.2 s.

If there is an odd number of values, the median is the middle value. In this case there is an even number of values so we take the middle two values, 8.0 s and 8.1 s, and calculate the median to be halfway between them, giving a result of 8.05 s.

The mean, median and mode are generally not the same values. Usually you can use the mean as an average in all your A-level Physics calculations.

B Guided question

Copy out the workings and complete the answers on a separate piece of paper.

1. A student measures the background count of radiation in the lab for five intervals of 5 minutes each and records the following total counts:

   200, 175, 215, 190, 210.

   a **Calculate the mean number of counts per 5 minutes.**

   Step 1: add the five values together.

   Step 2: divide the total by 5.

   b **Calculate the mean number of counts per second.**

   Step 1: calculate the number of seconds in 5 minutes.

   $5 \times 60 = 300$ seconds in 5 minutes

   Step 2: divide your answer from part **a** by 300.

C Practice question

2. A student uses an electronic timer to measure the time it takes for a ball bearing to fall 1.2 m ten times. The times obtained are:

   0.46 s, 0.44 s, 0.45 s, 0.46 s, 0.45 s, 0.44 s, 0.47 s, 0.46 s, 0.45 s, 0.47 s.

   Calculate the average value for the time of fall.

Probability

In A-level Physics, probability occurs in the context of radioactive decay. There is a certain probability per unit time that an unstable nucleus will decay.

In order to understand probability, consider spinning a coin: the probability that it will land heads up is 0.5 because there are just two possibilities. That does not mean that every two spins of the coin will give one heads and one tails. However, if you spin the coin 100 times, you will probably get approximately 50 heads and 50 tails, even though you could find the coin lands on heads (or tails) several times in a row during the 100 trials. If you spin the coin more times, you will find that, because the outcome of each spin is random, you are still unlikely to get half heads and half tails and, although there will be more outcomes that are different to the 'expected', the percentage difference between the expected value and the actual value will decrease the more often you spin the coin.

If you have three coins then the probability of obtaining heads on each of them for one spin is $0.5 \times 0.5 \times 0.5 = 0.125$ or $\frac{1}{8}$. Again, while you can expect to obtain three heads at one spin out of every eight spins, it does not mean that you will actually obtain three heads once every eight spins.

For a normal six-sided dice, the probability that it will land showing a 6 is one-sixth because for each throw there is one possibility out of six that the dice will land with a 6 upwards.

If you take 100 dice and throw them all, you can expect $\frac{100}{6} = 17$ to show a 6, although it is most unlikely that, if you try this, you will actually get 17 showing 6 because every throw is a random event.

If you remove all those dice showing a 6 and throw again, you would expect to find that $\frac{100-17}{6} = \frac{83}{6} \approx 14$ dice show a 6, although this is also uncertain since it is unlikely you actually had 17 showing 6 from the first throw and the outcome of the second throw is equally random.

The next time you throw you might expect about 12 to show 6 and so on.

However, these are random events and there will always be a difference between the expected outcome and the actual outcome. The percentage difference between the expected outcome and the actual outcome will decrease the more dice you throw and hence you will reduce your percentage error with a larger sample.

The number of dice landing on 6 at each throw depends on the number of dice left and the probability of throwing a 6. Plotting a graph of the expected number of dice left after each throw against number of throws gives the graph in Figure 3.1. If you try the experiment yourself, the data points will be scattered because these are random events — but the curve of best fit through the points will be similar to that shown. This is analogous to an exponential decay curve for radioactive decay, and you may have been shown or have done this experiment yourself as an analogy to radioactive decay. Radioactive decay is covered in more detail in Unit 7.

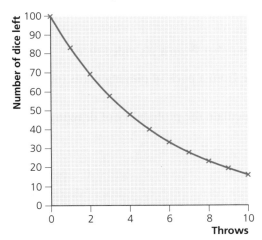

Figure 3.1 The theoretical number of dice left after each throw

B Guided questions

Copy out the workings and complete the answers on a separate piece of paper.

1 **Some games use an eight-sided dice. It has each of the numbers 1–4 on two different faces. Calculate the probability of the dice falling with 2 upwards when thrown.**

   Step 1: calculate the total number of possible outcomes when the dice is thrown.

   How many possibilities are there for the dice to fall with any number upwards?

Step 2: calculate the total number of favourable outcomes.

How many possibilities are there of the dice falling with 2 upwards?

Step 3: divide the number of favourable outcomes by the total number of possible outcomes.

The number of possibilities of the dice falling with 2 upwards divided by the total number of possibilities of the dice falling with any number upwards = _____

2 **A bag contains some coloured balls. Two are red, three are yellow and five are blue.**

A ball is taken from the bag at random.

Calculate the probability of a blue ball being picked.

Step 1: calculate the total number of possible outcomes.

The total number of balls = _____

Step 2: calculate the total number of favourable outcomes.

The total number of blue balls = _____

Step 3: divide the number of favourable outcomes by the total number of outcomes.

The probability of a blue ball being picked is the number of blue balls divided by the total number of balls = _____

C Practice questions

3 Calculate the probability of throwing an even number from a six-sided dice.

4 a Calculate the probability of drawing an ace from a pack of playing cards (not including the jokers).
   b The first five cards drawn are not aces. Calculate the probability of drawing an ace at the sixth attempt.

4 Algebra

Most of the relationships in A-level Physics can be described by an equation. It is therefore important that you can manipulate and re-arrange these equations. You may need to change the subject of the equation to find a value, for example, re-arranging $v = \frac{s}{t}$ to find a value for s given values of v and t, or you may need to re-arrange the equation so that you can draw a straight-line graph of two sets of values. In this simple case you could substitute the known values for v and t and solve the resulting equation, but not all equations will be as straightforward.

Some of the equations you will encounter are quadratic equations and you may need to be able to solve the equation to find a value. An example of this is the equation of motion $s = ut + \frac{1}{2}at^2$ where you may need to find a value of t given all the other values.

Re-arranging equations

The simple rule with equations is that to re-arrange an equation you must do the same to each side in order for both sides to remain equal.

You can:
- multiply or divide an equation by anything other than zero as long as you do the same to both sides of the equation
- add or subtract a quantity as long as you do the same to both sides of the equation
- square both sides of an equation (however, note that if $x = y + z$ then $x^2 = (y + z)^2$ not $x^2 = y^2 + z^2$)
- take the positive (or negative) square root of both sides of an equation
- find the reciprocal of both sides of an equation but take care: if $\frac{1}{x} = y + z$ then taking the reciprocal of both sides gives $x = \frac{1}{y + z}$ not $x = \frac{1}{y} + \frac{1}{z}$ (this error is commonly made when re-arranging equations for resistors in parallel)
- take a function, for example, log, sin, cos, of both sides. However, beware because if $x = y + z$, then $\sin x = \sin(y + z)$ not $\sin x = \sin y + \sin z$

Equations may have brackets in them, for example, $\varepsilon = I(r + R)$. In this case, the I multiplies both the r and the R so the equation becomes $\varepsilon = Ir + IR$.

The equation $\varepsilon = Ir + IR$ links the electromotive force (emf), ε, of a cell with the internal resistance, r, the load resistance, R, and the current flowing, I.

You may be asked to find r as the subject of the equation.

First try to get the part of the equation containing r on its own on one side. You can do this by subtracting IR from both sides: $\varepsilon - IR = Ir + IR - IR$. This gives $\varepsilon - IR = Ir$.

Now, to get r on its own as the subject of the equation, divide both sides by I: $\frac{\varepsilon - IR}{I} = \frac{Ir}{I} = r$.

Swap this round to write the final equation with r on the left as $r = \frac{(\varepsilon - IR)}{I}$.

As we shall see in the next part of this section, the brackets remind you that, when evaluating the equation by substituting values, you must evaluate the terms inside the brackets first.

Another example is re-arranging the equation for combining two resistors with values R_1 and R_2 in parallel to give an equivalent single resistor of total value R_T where $\frac{1}{R_T} = \frac{1}{R_1} + \frac{1}{R_2}$.

Here, R_T is on its own on one side of the equation but you cannot simply find the reciprocal of each individual term separately as that would make a different equation.

To re-arrange this equation, first combine the terms on the right-hand side (RHS) to give a single term. To do that you need to make the denominator the same for each term. (See Unit 2 for how to add fractions.)

Multiply the top and bottom of $\frac{1}{R_1}$ by R_2 to give $\frac{R_2}{R_1 \times R_2}$. Then multiply the top and bottom of $\frac{1}{R_2}$ by R_1 to give $\frac{R_1}{R_1 \times R_2}$. The expression then becomes $\frac{1}{R_T} = \frac{R_2}{R_1 \times R_2} + \frac{R_1}{R_1 \times R_2}$.

Since the denominators of the two fractions on the RHS are now the same they can be added to give $\frac{1}{R_T} = \frac{(R_1 + R_2)}{R_1 \times R_2}$. Only now can we take the reciprocal of both sides to give $R_T = \frac{R_1 \times R_2}{(R_1 + R_2)}$.

An example of an equation that includes squares is $hf = \phi + \frac{1}{2}mv^2$. This equation arises when studying the photoelectric effect.

Suppose you are required to find v as the subject of the equation. You cannot just square root each expression immediately, you first have to re-arrange to get v^2 on its own on one side, as the subject, and then find the square root.

First subtract ϕ from each side of the equation to give: $hf - \phi = \frac{1}{2}mv^2$.

Now divide both sides by $\frac{1}{2}m$ to give $\frac{hf - \phi}{1/2m} = v^2$. Finally take the square root of both sides to give $\sqrt{\frac{(hf - \phi)}{1/2m}} = v$.

A Worked examples

a Re-arrange the equation $pV = nRT$ to make T the subject of the equation.

Re-arrange so that T is on its own on one side.

Step 1: divide both sides by nR.

$$\frac{pV}{nR} = \frac{nRT}{nR}$$

Step 2: simplify. On the RHS, the values of nR cancel out leaving T, so $\frac{pV}{nR} = T$, and we have made T the subject of the equation.

Full worked solutions at **www.hoddereducation.co.uk/essentialmathsanswers**

b Make s the subject of the equation $v^2 = u^2 + 2as$.

Step 1: first re-arrange so that s is part of a single term on one side of the equation.

Subtract u^2 from both sides of the equation:

$v^2 - u^2 = u^2 - u^2 + 2as$

This gives $v^2 - u^2 = 2as$.

Step 2: divide both sides by $2a$.

$$\frac{v^2 - u^2}{2a} = \frac{2as}{2a}$$

Step 3: simplify. Cancel out $2a$ on the RHS to give $\frac{(v^2 - u^2)}{2a} = s$.

B Guided questions

Copy out the workings and complete the answers on a separate piece of paper.

1 **Kinetic energy is $E_k = \frac{1}{2}mv^2$.**

   Re-arrange the equation to make v the subject.

   Step 1: divide both sides by $\frac{1}{2}m$. (Remember that dividing by $\frac{1}{2}$ is the same as multiplying by 2.)

   Step 2: square root both sides to get v.

2 **The volume of a sphere is given by $V = \frac{4}{3}\pi r^3$.**

   Re-arrange this formula to give r as the subject.

   Step 1: re-arrange to get r^3 as the subject.

   Step 2: cube root, $\sqrt[3]{}$, both sides of the equation since $\sqrt[3]{r^3} = r$.

C Practice questions

3 Re-arrange the following equations, making the variable in brackets the subject of the equation.

   a $V = \dfrac{Q}{4\pi\varepsilon_0 r}$ (r)

   b $E = mc^2$ (c)

   c $E = \dfrac{Fl}{Ae}$ (e)

   d $P = I^2 R$ (I)

   e $\varepsilon = \dfrac{T_1 - T_2}{T_1}$ (T_1)

   f $s = \dfrac{u+v}{2}t$ (v)

   g $T = 2\pi\sqrt{\dfrac{l}{g}}$ (g)

h $E = \dfrac{Q}{4\pi\varepsilon_0 r^2}$ (r)

i $E = \dfrac{Q^2}{2C}$ (Q)

j (hard) $\gamma = \dfrac{1}{\sqrt{1 - v^2/c^2}}$ (v)

4 When two capacitors, C_1 and C_2, are connected in series, the total capacitance, C_T, is given by $\dfrac{1}{C_T} = \dfrac{1}{C_1} + \dfrac{1}{C_2}$.

A 100 μF capacitor is connected in series with a 150 μF capacitor.

Calculate the total capacitance of the combination.

Evaluating equations

The equations in the previous part of this section were all ones that you may encounter in your A-level Physics and you are likely to have to evaluate these equations by substituting values. The order in which you carry out the evaluation is important. Mathematicians use a mnemonic — BODMAS or BIDMAS — to summarise the order of carrying out operations.

B   stands for brackets; anything in brackets must be evaluated first.

O or I   stands for orders or indices. You should evaluate anything with an index such as 5^2 or $\sqrt{5}$.

D M   stands for division and multiplication; these operations should be carried out next.

A S   stands for addition and subtraction; finally carry out any additions or subtractions which should be evaluated starting from the left.

As an example, evaluate $7 \times 3 - 8 \div 4 + 6 \times (5^2 - 4^2)$.
- First evaluate the brackets $(5^2 - 4^2)$ and within the brackets use the BODMAS order, so $5^2 = 25$ and $4^2 = 16$ therefore $(25 - 16) = 9$.
- Now carry out the multiplications and divisions.
  $7 \times 3 = 21$, $8 \div 4 = 2$, $6 \times 9 = 54$
- Finally carry out the additions and subtractions.
  $21 - 2 + 54 = 73$

Be careful as sometimes brackets are omitted. For example, $\dfrac{16 - 4}{3}$ is understood to mean $\dfrac{(16 - 4)}{3}$ so you should evaluate the brackets first $(16 - 4) = 12$ before dividing by 3 to give 4.

> **TIP**
>
> When writing your calculations, always put brackets around quantities that must be evaluated first. This will help you when you enter the calculation into your calculator as it is easy to miss out a set of brackets. The calculator will carry out the calculation exactly as you have entered it, using the rules above.

A Worked examples

a Evaluate $\frac{(2s - ut)}{t} = v$ given that $s = 10\,\text{m}$, $u = 2\,\text{m s}^{-1}$ and $t = 0.5\,\text{s}$.

Step 1: remember that the top line is taken to have brackets around it, i.e. $\frac{(2s - ut)}{t} = v$.

Step 2: substitute the numbers.

$$\frac{(2 \times 10 - 2 \times 0.5)}{0.5} = v$$

Step 3: using the BODMAS order, evaluate the brackets.

$(20 - 1) = 19$

Step 4: complete the evaluation by dividing.

$19 \div 0.5 = 38\,\text{m s}^{-1}$

b Evaluate $v = \sqrt{c^2\left(1 - \frac{1}{\gamma^2}\right)}$ given that $c = 3.0 \times 10^8\,\text{m s}^{-1}$ and $\gamma = 2.0$.

Step 1: using the BODMAS order, evaluate the brackets.

$$\left(1 - \frac{1}{\gamma^2}\right) = \left(1 - \frac{1}{2^2}\right) = (1 - 0.25) = 0.75$$

Step 2: evaluate the terms inside the square root sign (taken to be inside brackets).

$c^2 \times 0.75 = (3.0 \times 10^8)^2 \times 0.75 = 6.75 \times 10^{16}$

Step 3: take the square root.

$v = \sqrt{6.75 \times 10^{16}} = 2.6 \times 10^8\,\text{m s}^{-1}$

B Guided questions

Copy out the workings and complete the answers on a separate piece of paper.

1 **Evaluate the following.**

  a $4 + 3 \times 5$

  Use the BODMAS order: multiplication first, then addition.

  b $(3 + 4) \times 5$

  Using the BODMAS order, evaluate the brackets first then multiply.

  c $\frac{25 + 15}{4}$

  The top is taken to have brackets around it so evaluate $(25 + 15) = 40$ first.

  d $6 \times 15 + (26 - 3^2)$

  Step 1: using the BODMAS order, evaluate the brackets first, and within the brackets use the order of index then subtraction to give 17.

  Step 2: multiply next. $6 \times 15 = 90$

  Step 3: finally add.

2 Evaluate $\dfrac{-8 + \sqrt{8^2 - 4 \times 3 \times 2}}{2 \times 3}$

Step 1: using the BODMAS order, evaluate the part inside the square root sign (taken to be inside brackets).

$\sqrt{8^2 - 4 \times 3 \times 2} = 6.325$

Step 2: evaluate the top of the fraction (also taken to be inside brackets).

$-8 + 6.325 = -1.675$

Step 3: evaluate the bottom of the fraction (also taken to be inside brackets).

$2 \times 3 = 6$

Step 4: divide –1.675 by 6 to give the final value.

C Practice questions

3 Evaluate the following:

a $7 - \dfrac{(5 \times 3^2)}{9}$

b $13 \times 5^2 + \dfrac{54}{6}$

c $\sqrt{3^2 + \dfrac{4}{3} \times 12 + 11}$

d $\dfrac{100 \times 50}{100 + 50}$

e $\dfrac{4 + \sqrt{(-4)^2 - (4 \times 8 \times -24)}}{2 \times 8}$

4 Evaluate $g = \dfrac{4\pi^2 l}{T^2}$ where l is 2.5 m and T is 3.17 s.

5 Evaluate $v = \sqrt{\dfrac{2(hf - \phi)}{m}}$ where $h = 6.6 \times 10^{-34}$ J s, $f = 8.0 \times 10^{14}$ Hz, $\phi = 3.4 \times 10^{-19}$ and $m = 9.1 \times 10^{-31}$ kg.

Quadratic equations

A quadratic equation is one where the unknown quantity, x, is in the equation as x^2 and often as x too. A general quadratic equation is of the form $ax^2 + bx + c = 0$, where a, b and c are constants.

The general formula for calculating the value of x is $x = \dfrac{-b \pm \sqrt{b^2 - 4ac}}{2a}$, remembering to take account of any negative signs in the equation. The $\pm$ comes in because there is both a positive and negative square root of any positive number, and both possibilities have to be calculated.

The quadratic equation you are most likely to be asked to solve is one of the equations of motion: $s = ut + \frac{1}{2}at^2$ where u is the initial velocity, a is the acceleration and s is the displacement at time t.

If we re-arrange this equation to be in the same form as the general quadratic equation we would write it as $\frac{1}{2}at^2 + ut - s = 0$, where t is the unknown. When specific values are substituted for a, u and s, this equation can be compared with the general quadratic equation and solved using the formula.

Here you may be given the height, s, from which a ball is dropped and asked to calculate the time, t, taken to hit the ground. You may obtain both a positive and a negative value for t but for this particular quadratic, you can usually ignore the negative value because it has no meaning in this particular context. The worked examples show how this equation is used and solved.

 Worked examples

a **A ball is dropped from a height of 6.0 m to the ground.**

g is 9.8 N kg^{-1}.

Calculate the time taken for the ball to reach the ground.

Since $u = 0$, in this case, we don't actually need the formula for solving a quadratic equation, but we will use it to show the process and then try an example when u is not zero.

Method 1

Step 1: re-arrange the equation $s = ut + \frac{1}{2}at^2$ to be in the form of the general equation for a quadratic.

Subtract s from both sides.

$$\frac{1}{2}at^2 + ut - s = 0$$

Step 2: substitute the values you know.

The numerical value of the acceleration, a, is equal to g, 9.8.

$$\frac{1}{2} \times 9.8 \times t^2 + 0 \times t - 6.0 = 0 \Rightarrow 4.9t^2 + 0t - 6.0 = 0$$

Step 3: use the formula $x = \frac{-b \pm \sqrt{b^2 - 4ac}}{2a}$ to solve the quadratic equation.

Comparing with the general quadratic equation, $a = 4.9$, $b = 0$, $c = 6.0$.

$$t = \frac{-0 \pm \sqrt{0^2 - 4 \times 4.9 \times -6.0}}{2 \times 4.9} = \frac{\pm 10.84}{9.8} = \pm 1.1\,\text{s}$$

Clearly the negative time is not relevant in this case, so the ball takes 1.1 s to reach the ground.

Method 2

This simpler method can be used because $u = 0$, so, in this case, it isn't necessary to use the quadratic formula. Remember that $0 \times t = 0$.

The original equation simplifies to $6.0 = 4.9 \times t^2 \Rightarrow t = \sqrt{\frac{6.0}{4.9}} = 1.1\,\text{s}$.

b **The same ball as in worked example a is now thrown downwards with a velocity of $2\,\text{m}\,\text{s}^{-1}$. Calculate the time taken for the ball to reach the ground.**

Since u is in the same direction (downwards) as the acceleration, a, it must have the same sign.

It doesn't matter whether we decide downwards is positive or negative, as long as we are consistent and use the same sign for both u and a.

This time we do need to use the quadratic formula.

Step 1: using $\frac{1}{2}at^2 + ut - s = 0$, substitute the values you know.

$$\tfrac{1}{2} \times 9.8 \times t^2 + 2.0 \times t - 6 = 0$$

Step 2: use the formula to solve the quadratic equation.

Comparing with the general quadratic equation, $a = 4.9$, $b = 2$, $c = 6.0$.

$$t = \frac{-2 \pm \sqrt{2^2 - 4 \times 4.9 \times -6.0}}{2 \times 4.9}$$

$$= \frac{-2 \pm \sqrt{121.6}}{9.8} = \frac{-2 \pm 11.03}{9.8} = \frac{9.03}{9.8} = 0.92\,\text{s} \text{ or } \frac{-13.03}{9.8} = -1.3\,\text{s}$$

The negative time does not have any meaning so is ignored.

The ball takes 0.92 s to reach the ground.

c **Suppose now that the ball is first thrown upwards, rather than downwards, with a velocity of $2\,\text{m}\,\text{s}^{-1}$ from a height of 6.0 m. Calculate the time taken to reach the ground.**

u is now $-2.0\,\text{m}\,\text{s}^{-1}$ because the velocity is now upwards and in the opposite direction to the direction of the acceleration. This is shown by the negative sign for u.

Step 1: using $\frac{1}{2}at^2 + ut - s = 0$, substitute the values you know.

$$\tfrac{1}{2} \times 9.8 \times t^2 - 2.0 \times t - 6 = 0$$

Step 2: use the formula to solve the quadratic equation.

$$t = \frac{-(-2) \pm \sqrt{(-2)^2 - 4 \times 4.9 \times -6.0}}{2 \times 4.9}$$

$$= \frac{2 \pm \sqrt{121.6}}{9.8} = \frac{2 \pm 11.03}{9.8} = \frac{-9.03}{9.8} = -0.92\,\text{s} \text{ or } \frac{+13.03}{9.8} = +1.3\,\text{s}$$

Again, the negative value of time is not relevant.

It takes 1.3 s for the ball to reach the ground.

You will see from the worked examples that when using the quadratic formula it is important to get the signs correct in the equation and to use them correctly when solving it.

> **TIP**
>
> If the quadratic equation does not have a term in x but only in x^2 then you do not need to use the quadratic formula to find a value for x.

Full worked solutions at www.hoddereducation.co.uk/essentialmathsanswers

B Guided question

Copy out the workings and complete the answers on a separate piece of paper.

1 A car travelling at 8.0 m s⁻¹ starts to accelerate at 0.8 m s⁻².

   Calculate how long the car takes to cover 210 m.

   Step 1: re-arrange the equation $s = ut + \frac{1}{2}at^2$ to be in the form of the general equation for a quadratic, $ax^2 + bx + c = 0$.

   Step 2: substitute the values you know.

   $s = 210\,\text{m}, u = 8.0\,\text{m s}^{-1}, a = 0.8\,\text{m s}^{-2}$

   Step 3: use the formula to solve the quadratic equation.

   $t =$ _____

C Practice questions

2 Find the value of x for each of the following:
   a $30x^2 + 8x - 6 = 0$
   b $2x^2 + 5x - 63 = 0$
   c $4.9x^2 + 16x - 32 = 0$

3 A ball is thrown directly downwards from the top of a 200 m cliff at 5.0 m s⁻¹.

   g is 9.8 N kg⁻¹.

   Calculate the time taken for the ball to hit the ground at the bottom of the cliff.

5 Graphs

Graphs are an essential tool for a physicist; a good graph is able to display a trend or a pattern clearly and provides the opportunity for further analysis. However, in most cases where further analysis is required, it is best to plot the data in a form where a straight-line graph is expected.

Straight lines

The standard equation of a straight line is $y = mx + c$, where m is the gradient and c is the intercept on the y-axis. This is the value of y when x is zero. (See Figure 5.1.) However, this will only be the intercept shown on the y-axis of the graph if the graph is drawn so that the x-axis starts at zero. If your data is such that you decide to start the x-axis not at zero, then any intercept you measure from the graph will be incorrect. You can decide not to start the y-axis at zero as long as the intercept is clearly visible.

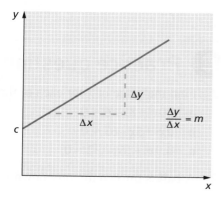

Figure 5.1 A straight-line graph with equation $y = mx + c$

The gradient of a graph is always found by measuring the increase on the y-axis, Δy, and dividing by the corresponding change in the x-axis, Δx. Remember to work out the distance using the units of the axis, not by counting the squares on the graph paper.

A Worked example

Figure 5.2 shows a graph of distance against time for a toy car.

   i Explain the significance of the gradient of the graph.

   The gradient of the graph is the speed, $v = \frac{\Delta s}{\Delta t}$.

   ii Calculate the gradient of the graph.

   $m = \frac{4.5 - 2.0}{10 - 0} = \frac{2.5 \text{ cm}}{10 \text{ s}} = 0.25 \text{ cm s}^{-1}$.

   iii What does the intercept on the y-axis indicate?

   The intercept, c, is 2.0 cm. This means that at the starting time, $t = 0$, the car was 2.0 cm from the zero position.

   iv Write down the equation for the line.

   The equation of this graph would be $s = 0.25 \times t + 2.0$.

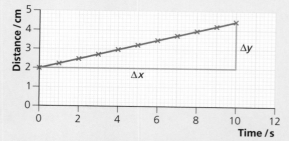

Figure 5.2 A straight-line graph of distance, s, against time, t

As well as being asked to find a gradient and an intercept of a straight-line graph as in the example, you might use a straight-line graph to test a hypothesis or to confirm a pattern of data that is expected or guessed.

For example, a mass hung on a spring oscillates in a vertical direction with a periodic time, T, which depends on the mass, M, and the spring constant, k. If the mass is changed, the resulting periodic time is measured and a graph plotted of T against M (by convention, T is plotted on the y-axis and M on the x-axis), then a curve is obtained as in Figure 5.3.

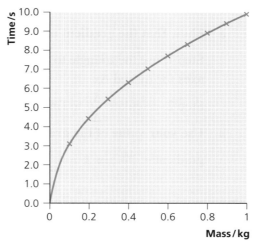

Figure 5.3 Time against mass for a mass–spring system

This graph tells us little other than that the relationship is non-linear. To be useful the graph needs to be a straight line. In this case, we may know that the expected relationship is $T = 2\pi\sqrt{\frac{M}{k}}$. This does not look much like the equation for a straight line, but by squaring both sides we get $T^2 = 4\pi^2 \frac{M}{k}$. And, with a small re-arrangement, we get $T^2 = \frac{4\pi^2}{k} M$.

Comparing this with the equation for a straight line, $y = mx + c$, it can be seen that T^2 is the 'y' variable, M is the 'x' variable and the gradient, m, is $\frac{4\pi^2}{k}$. So, rather than plotting T against M, we should plot T^2 against M, which will give a straight line as shown in Figure 5.4.

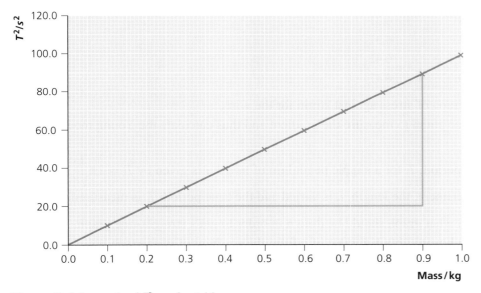

Figure 5.4 A graph of T^2 against M

By plotting this as a straight line, we are able to measure the gradient and to find a value for k, which we may not have known before analysing the results of the experiment.

Almost every equation you will encounter at A-level can be re-arranged in this way in order to be able to plot a straight line using experimental data.

B Guided question

Copy out the workings and complete the answers on a separate piece of paper.

1 a Show that the gradient of the graph shown in Figure 5.4 is about $100\,\text{s}^2\,\text{kg}^{-1}$.

   Step 1: use the triangle to find Δy and Δx. Do not forget to use the units of the axes for Δy and Δx.

   Step 2: calculate $\dfrac{\Delta y}{\Delta x}$ to find the gradient.

   b Use your value of the gradient to find a value for k.

   gradient $= \dfrac{4\pi^2}{k} \Rightarrow k = \dfrac{4\pi^2}{\text{gradient}} = $ _____

C Practice questions

2 The Young modulus, E, of a wire is given by $\dfrac{\text{tensile stress}}{\text{tensile strain}}$.

   Values of stress and strain for a copper wire are shown in Table 5.1.

   Table 5.1

Strain	Stress/GPa
0.0000	0.00
0.0004	0.05
0.0008	0.09
0.0012	0.14
0.0016	0.19
0.0020	0.23
0.0024	0.28
0.0028	0.33
0.0032	0.37
0.0036	0.42
0.0040	0.47

   a Plot a graph of stress against strain for the wire.

   b Use your graph to measure the gradient and thus calculate the Young modulus for copper.

3 The time period, T, for a simple pendulum of length l is given by $T = 2\pi\sqrt{\dfrac{l}{g}}$.

   In an experiment, the length, l, is varied and the time period, T, is measured.

   a Explain what graph should be drawn in order to get a straight-line graph.

   b Explain how to obtain a value of g from the gradient of your graph.

Shapes of graphs for different functions

Although straight-line graphs are the most useful for analysis, there are a number of other graphs that occur in A-level Physics that you should be able to identify and, if necessary, sketch.

Boyle's law states that the pressure, p, of an ideal gas is related to the volume, V, by $p \propto \dfrac{1}{V}$. A graph of volume against pressure for an ideal gas is shown in Figure 5.5. This is an example of a graph of the type $y = \dfrac{k}{x}$.

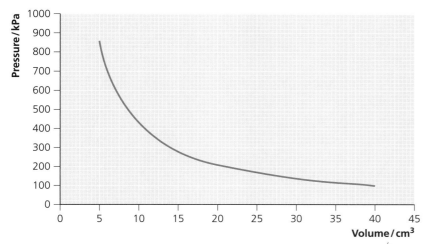

Figure 5.5 Boyle's law for an ideal gas, a graph of the type $y = \frac{k}{x}$

The kinetic energy of a body depends on the square of its velocity. A graph of kinetic energy against velocity is shown in Figure 5.6. This is an example of a graph of the type $y = kx^2$.

The force, F, between two masses, M and m, a distance r apart is given by the expression $F = \frac{GMm}{r^2}$. A graph of the force between the Earth and a rocket is shown in Figure 5.7. This is an example of a graph of the type $y = \frac{k}{x^2}$. Note that the line never reaches either axis.

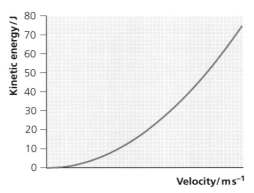

Figure 5.6 Kinetic energy of a moving body, a graph of the type $y = kx^2$

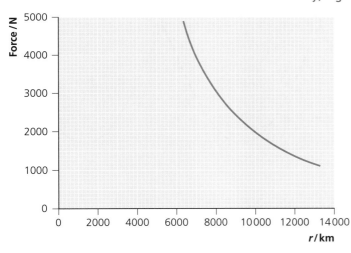

Figure 5.7 The force between the Earth and a 500 kg rocket, a graph of the type $y = \frac{k}{x^2}$

The displacement, s, of a simple harmonic oscillator which is released from its maximum displacement, A, at time $t = 0$ is given by the expression $s = A\cos(\omega t)$ where ω is a constant.

A graph of displacement against time is shown in Figure 5.8. In this case, since the maximum and minimum values of $\cos(\omega t)$ are ± 1, the maximum and minimum values of s are $\pm A$.

This is an example of a graph of the type $y = \cos x$.

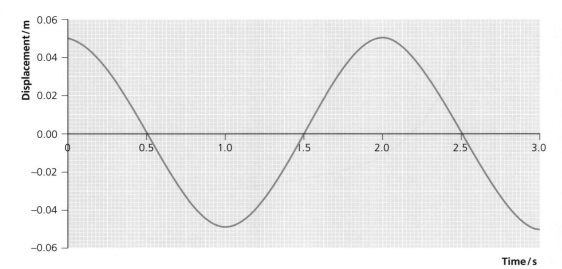

Figure 5.8 The displacement of an oscillator with time, a graph of the type $y = \cos x$

The velocity of the same simple harmonic oscillator is given by the expression $v = -A\omega \sin(\omega t)$ and a graph showing this is given in red in Figure 5.9. The negative sign in the expression means that the graph is inverted. A graph of the type $y = \sin x$ would be the same shape but inverted as shown by the dashed green curve in Figure 5.9.

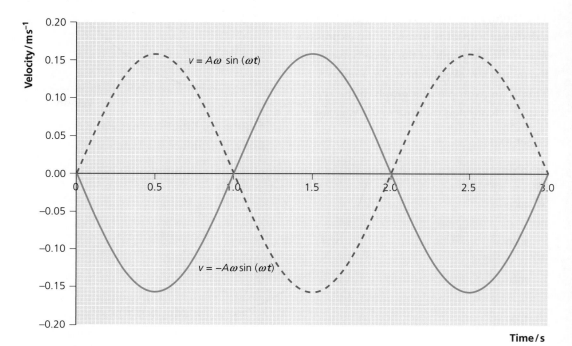

Figure 5.9 The velocity of an oscillator with time

The kinetic energy of an oscillator is given by $E_k = \frac{1}{2} mA^2\omega^2 \sin^2(\omega t)$ and the potential energy of an oscillator is given by $E_p = \frac{1}{2} mA^2\omega^2 \cos^2(\omega t)$ where m is the mass of the oscillator. These graphs, as they are $\sin^2 x$ and $\cos^2 x$, cannot be negative and so are all above the x-axis. They are of the type $y = \sin^2 x$ and $y = \cos^2 x$ respectively, and examples of these types of graph are in Figure 5.10. They are for the same oscillator as in Figures 5.8 and 5.9 so you can see how graphs of $\sin x$ and $\sin^2 x$ compare, as well as $\cos x$ and $\cos^2 x$.

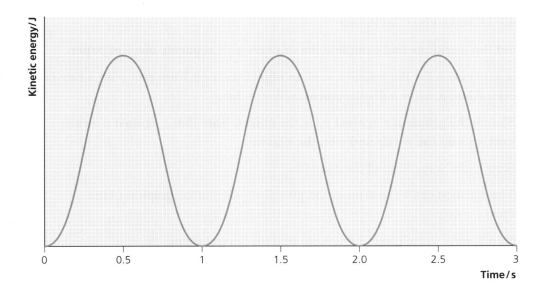

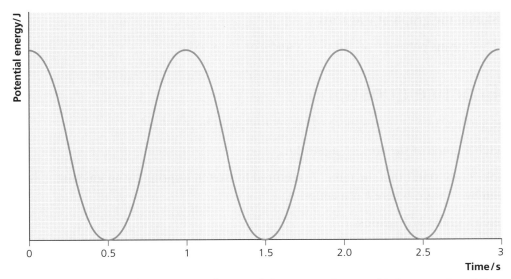

Figure 5.10 The kinetic energy and potential energy of an oscillator, graphs of the type $y = \sin^2 x$ (top) and $y = \cos^2 x$ (bottom)

> **TIP**
>
> The graphs shown give real examples of the shapes of curves. You should be familiar with the general shape of each of the types of graph.

> **A** Worked example
>
> **The expression for the de Broglie wavelength, λ, of an electron is given by $\lambda = \dfrac{h}{mv}$ where h and m are constant and v is the speed of the electron.**
>
> **Sketch a graph of λ against v for an electron.**
>
> Step 1: check the form of the equation.
>
> This equation is of the form $y = \dfrac{k}{x}$.
>
> Step 2: refer to the general graphs to find the correct shape.
>
> The graph is similar in shape to Figure 5.5 with λ on the y-axis and v on the x-axis.

B Guided questions

Copy out the workings and complete the answers on a separate piece of paper.

1. The expression for the velocity, v, of an object starting from rest with a constant acceleration, a, after a displacement, s, is $v^2 = 2as$.

    a Sketch the graph of s against v for the object. Note that s against v means putting s on the y-axis and v on the x-axis.

    Step 1: check the form of the equation.

    The equation can be re-arranged to give $s = \frac{1}{2a}v^2$, so the graph is of the form $y = kx^2$.

    Step 2: refer to the general graphs to find the correct shape.

    Figure 5.6 gives the general shape of the graph. There are no figures to put on the axes but you must label the axes.

    b Sketch the graph of v against s.

    This means putting v on the y-axis and s on the x-axis.

C Practice questions

2. The surface area, A, of a sphere of radius r is given by $A = 4\pi r^2$.

    Sketch a graph of the area, A, against r.

3. Hooke's law states that the force, F, required to stretch a spring by an amount, x, is given by $F = kx$, where k is a constant.

    Sketch a graph of F against x.

4. The force, F, between two charges a distance r apart is given by $F = \frac{kQq}{r^2}$, where k, Q and q are constant.

    Sketch a graph of F against r.

5. The potential, V, at a distance r from a charge, $-Q$, is given by $V = \frac{kQ}{r}$, where k is a constant.

    Sketch a graph of V against r.

TIP

When asked to sketch (rather than draw) a graph, you do not need to work out a set of data to plot points. But you must label axes and if any data is given, then use the data values to label any significant points, for example, an amplitude or a time period of an oscillation.

Rates of change

Sometimes it is not possible to draw a straight-line graph from a set of data yet we may need to find the rate at which the quantity plotted on the *y*-axis varies with the quantity plotted on the *x*-axis (which is often time).

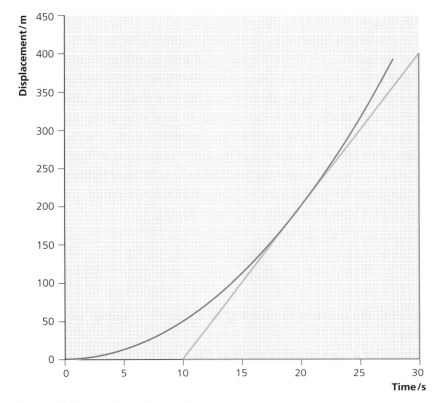

Figure 5.11 Accelerating motion

Consider the motion shown in Figure 5.11. The graph shows the displacement, *s*, of a body against time, *t*. We can measure two different values for the velocity. After 20 s, the body has moved 200 m and therefore the average velocity over 20 s is $\frac{s}{t} = \frac{200}{20} = 10$ m s^{-1}. However, the actual velocity at 20 s, known as the instantaneous velocity or the rate of change of displacement with time, $\frac{\Delta s}{\Delta t}$, is given by the gradient of the curve at 20 s. This is found by drawing a tangent to the curve at 20 s, as shown by the green line in Figure 5.11, and measuring the gradient of the tangent using the triangle shown in blue in Figure 5.11.

The tangent to the curve is a line at the same slope or gradient as the curve at that point. On a smooth curve this line will usually be symmetrical about the point for which the tangent is required and this has to be judged mainly by eye. One way of drawing a tangent is to use a clear ruler and work from the outside of the curve, using the markings on the ruler and the grid of the graph paper to try to check for symmetry. The tangent should just touch the curve at the point at which the gradient is being measured. Always draw the longest line that will fit on your graph paper so that you can use the largest triangle to find Δy and Δx. If you practise on different graphs, you should get better at drawing an accurate tangent. However, it is not a precise method and will only ever give an approximate answer.

In Figure 5.11, the gradient of the tangent at 20 s is given by

$$\frac{\Delta s}{\Delta t} = \frac{400}{30-10} = \frac{400}{20} = 20 \text{ m s}^{-1}$$

If the graph had been a straight line, the average and the instantaneous velocity would have had the same value.

Mathematically, finding the gradient of any graph is done by using a branch of mathematics called calculus. The differential of a function is the process of finding the gradient or rate of change of the graph at any point. This process is indicated by the differential operator, $\frac{dy}{dx}$. By finding the tangent to the curve of displacement against time and measuring its gradient, we have estimated a value for $\frac{\Delta s}{\Delta t}$ for this curve at the time of 20 s. By making the values of Δs and Δt smaller and smaller (we say 'as Δt tends to zero'), the value obtained for the gradient becomes closer and closer to the exact instantaneous value. This is the process of differentiation. If we know the equation of the curve then, using differentiation, we can obtain a second equation which allows the gradient to be calculated rather than measured at any point. However, for your A-level Physics, you are not required to do this; you are required only to draw a tangent and measure its gradient.

A Worked example

a **Show that the gradient of the tangent to the curve in Figure 5.11 at a time of 10 s gives an instantaneous velocity of about 10 m s⁻¹.**

Step 1: make a copy of Figure 5.11 and draw a tangent to the curve at 10 s, making sure it is symmetrical to the curve.

A tangent to the curve at 10 s might go from the point (5, 0) to the point (20, 150).

Step 2: find the gradient of the tangent, which represents the instantaneous velocity.

$$\text{instantaneous velocity} = \text{gradient of tangent} = \frac{150-0}{20-5} = \frac{150}{15} = 10 \text{ m s}^{-1}$$

b **Calculate the average velocity after a time of 10 s.**

Measure the displacement at time 10 s and divide by the time.

$$\text{average velocity} = v = \frac{s}{t} = \frac{50}{10} = 5 \text{ m s}^{-1}.$$

c **Calculate the change in instantaneous velocity, Δv, between 10 s and 20 s using the answer to worked example a and the information in the text.**

$$\text{change in instantaneous velocity } \Delta v = (20-10) = 10 \text{ m s}^{-1}$$

d **Calculate the acceleration of the object.**

$$\text{acceleration} = \frac{\Delta v}{\Delta t} = \frac{10}{10} = 1.0 \text{ m s}^{-2}$$

B Guided question

Copy out the workings and complete the answers on a separate piece of paper.

1 Look at the motion shown in Figure 5.12.

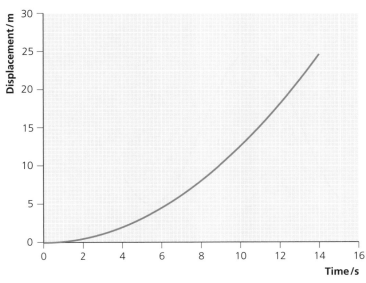

Figure 5.12

a **Calculate the average velocity after 6.0 s.**

The average velocity is given by displacement/time.

Measure the displacement at 6.0 s and divide by the time, 6.0 s.

average velocity = _____

b **Calculate the instantaneous velocity at 6.0 s.**

Step 1: draw a tangent to the curve at 6.0 s, making sure it is symmetrical to the curve.

Your tangent might pass through points (3, 0) and (12, 13.5).

Step 2: find the gradient of the tangent, which represents the instantaneous velocity.

instantaneous velocity = _____

C Practice questions

2 In the motion shown in Figure 5.12, find the instantaneous velocity at time $t = 10$ s.

3 Figure 5.13 shows the displacement of an oscillator with time.

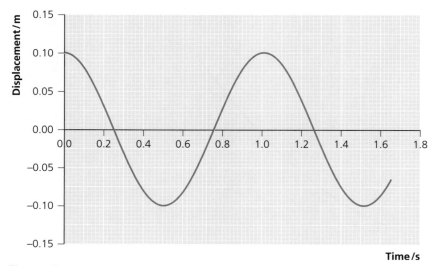

Figure 5.13

a By drawing a tangent to the curve at the first point where the displacement is zero, on a copy of the graph, calculate the velocity at this point.

b Draw a tangent to the curve at time $t = 0.8$ s and measure the gradient to find the instantaneous velocity of the oscillator at this time.

c Explain why the tangent to the curve at $t = 1.0$ s shows that the instantaneous velocity of the oscillator is zero at this point.

Area under a graph

The area under a graph may be of significance. For example, consider the graph of velocity against time in Figure 5.14, which shows a constant velocity of $10 \, \text{m s}^{-1}$ for 10 s.

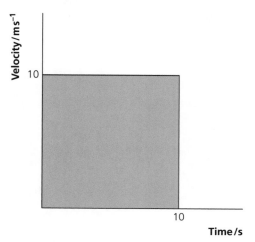

Figure 5.14

Displacement is always $v \times t$ and this can be seen to be equal to the shaded area in Figure 5.14, 100 m.

Now consider the motion shown in Figure 5.15. Here the velocity is not constant but is increasing at a uniform rate; there is a constant acceleration.

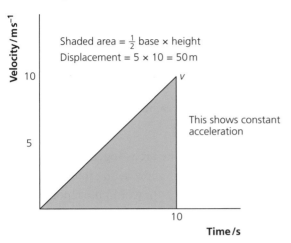

Figure 5.15

The displacement is the average velocity × time, which is $\frac{10}{2} \times 10 = 50$ m. This is the area of the shaded triangle, $\frac{1}{2} \times$ height × base.

In general, the displacement is always the area under a velocity–time graph.

Now consider a different situation as shown in Figure 5.16.

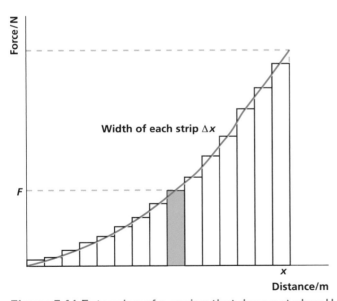

Figure 5.16 Extension of a spring that does not obey Hooke's law

Figure 5.16 is a graph of force against extension for a spring that does not obey Hooke's law, but gets stiffer the more it is stretched. The spring is being stretched by an amount x.

The work done when a force F moves through a small distance Δx is $F \times \Delta x$. This is the area of the shaded strip in Figure 5.16. The total work done in stretching the spring a distance x is given by the area under the curve, which is the sum of the areas of all the strips of width Δx.

We would write that the work done $W \approx \sum_0^x F\Delta x$, which, in words, means that the total work done is approximately equal to the sum of all the strips $F\Delta x$ between zero and x.

However, as can be seen, this is not quite correct because each strip $F \times \Delta x$ is a rectangle and therefore does not fit under the curve exactly. The narrower we make each strip, the more strips we will have and the closer the total area of all the strips will be to the true area.

We might therefore write that $W = \lim_{\Delta x \to 0} \sum_0^x F\Delta x$. This means that the value of the sum becomes closer to being exactly W as the width of each strip gets closer to zero ($\lim_{\Delta x \to 0}$ means in the limit as Δx tends to zero), and the narrower the strips, the more strips there are.

Mathematically, taking the area of narrow strips and adding the areas together is a process of calculus called integration. The area under a line or curve is the integral of the equation for the line or curve and, for the curve in Figure 5.16, would be denoted by the symbol $\int_0^x F\, dx$. For A-level Physics, you are not required to carry out the process of integration but you should recognise the term and, more importantly, realise that the area under a line or curve can be significant.

Work done, or energy stored, as the area under a graph is frequently used. When the graph is not a straight line, the area has to be measured in other ways and this is usually done by counting the squares under the curve.

To do this, you need to calculate the work done that is represented by one square. How big a square you use is up to you but normally a $10\,\text{mm} \times 10\,\text{mm}$ square on standard graph paper is suitable.

To calculate the work done, ΔW, represented by one square, measure ΔF (in N) represented by the vertical side of the square, making sure you take note of the units of the axis. Then measure Δx (in m) represented by the horizontal side of the square, again making sure you note the units of the axis. Then multiply $\Delta F \times \Delta x$ to find ΔW.

Now count the squares under the curve. The rule is to count any squares that are half or more than half as one and any squares that are less than half as zero. This may sound imprecise but it works surprisingly well.

Finally, having counted the squares, multiply the number of squares by ΔW to give the total work done, W. The following worked example goes through this process for you.

A Worked example

Figure 5.17 shows the graph of force against extension for a material that does not obey Hooke's law.

Calculate the work done or energy stored when the material is extended to 10 cm.

The work done is the area under the graph between the curve and the *x*-axis up to the value of $x = 10\,\text{cm} = 0.1\,\text{m}$.

Step 1: calculate the work done represented by one large square.

Use one large square, which is
$0.01\,\text{N} \times 0.01\,\text{m} = 1.0 \times 10^{-4}\,\text{J}$.

Step 2: count the number of squares under the graph, using the rule given in the text.

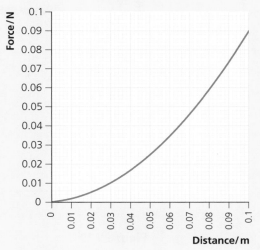

Figure 5.17

Counting squares that are half or more than half as one square, and squares that are less than half as zero, gives a total of 31 squares.

Step 3: multiply the number of squares by the work done per square.

The work done or energy stored is $31 \times 1.0 \times 10^{-4}\,\text{J} = 3.1 \times 10^{-3}\,\text{J}$.

This method of counting squares is surprisingly accurate. A value for the work done found from integration gives the work done as $3.2 \times 10^{-3}\,\text{J}$, an error of only about 3%.

B Guided questions

Copy out the workings and complete the answers on a separate piece of paper.

1 Figure 5.18 is a graph of charge, Q, against potential difference, V, for a capacitor. Energy $= V \times Q$.

   Calculate the energy stored in the capacitor when it is charged to 10 V.

   - The energy stored is the area under the graph, which is the area of the triangle under the graph up to 10 V.
   - You must make sure you use the units of the axes so the area of a triangle $A = \frac{1}{2} \times \text{base} \times \text{height}$.
   - The base of the triangle is 10 V and the height of the triangle is 0.030 C.

   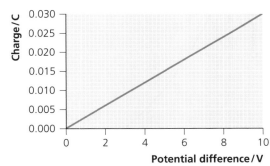

   Figure 5.18

   Energy stored = _____

2   Figure 5.19 shows the force on 1 kg at various distances from the Earth.

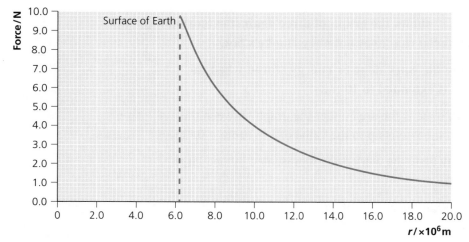

Figure 5.19

Use the graph to calculate the energy lost by a 1 kg mass when it is brought from a distance of 20×10^6 m to the surface of the Earth.

The question is asking you to calculate the area under the graph up to the line labelled 'Surface of Earth'.

Step 1: calculate the work done represented by one large square.

Step 2: count the number of squares under the graph, using the rule given in the text.

Step 3: multiply the number of squares by the work done per square.

Work done = _____

C Practice questions

3   Figure 5.20 shows the graph of current against time for a capacitor discharging.

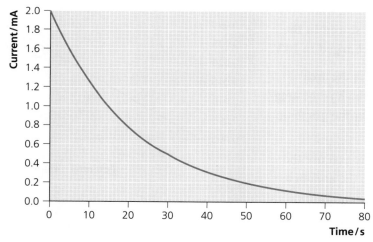

Figure 5.20

The charge stored on the capacitor is given by the area under the graph.

Calculate the area and thus the charge stored on the capacitor.

4 Figure 5.21 shows the graph of rate of flow of water from a burette with time.

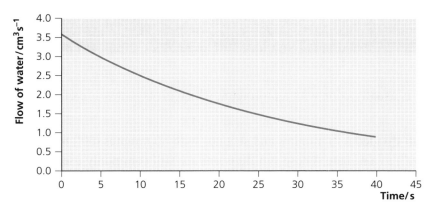

Figure 5.21

The volume of water that has drained from the burette is given by the area under the graph.

Calculate the volume of water that has drained from the burette in 40 s.

6 Geometry and trigonometry

Measuring angles is usually done with a protractor, which gives a measure of an angle in degrees.

Degrees can be sub-divided into minutes and seconds with 60 minutes (60′) to a degree and 60 seconds (60″) to a minute. These sub-divisions are still commonly used in astronomy, although for most calculations we now use decimal degrees.

For example: 25° 16′ 35″ would be written in decimal degrees as $25 + \frac{16}{60} + \frac{35}{3600} = 25.2764°$.

Before starting A-level, it is quite likely that you will have measured angles only in degrees. In physics, at A-level and beyond, a different measurement of angle is preferred. This is measurement of angles in radians.

Radians

Imagine a circle as shown in Figure 6.1. Radian measure is linked to the angle subtended at the centre of a circle by two radii. The angle θ in radians is defined as:

$$\theta = \frac{\text{arc length, } s}{\text{radius, } r} = \frac{s}{r}$$

If the length of the arc is equal to the radius of the circle, $s = r$, then

$$\theta = \frac{s}{r} = 1 \text{ radian}$$

If we have a complete rotation, as shown in Figure 6.2, then the arc length, s, is $2\pi r$ (the circumference of the circle) and the angle at the centre, θ, in radians is given by:

$$\theta = \frac{2\pi r}{r} = 2\pi$$

So a complete rotation is 2π radians.

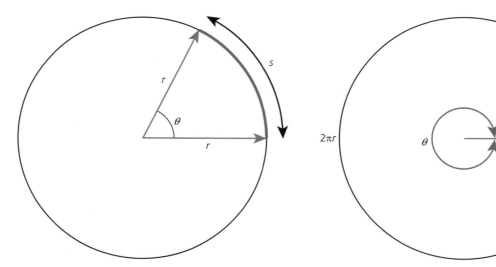

Figure 6.1 Angle in radians

Figure 6.2 A complete rotation

Degrees and radians

A complete rotation, 360°, is equivalent to 2π radians.

180° is equivalent to π radians.

90° is equivalent to $\frac{\pi}{2}$ radians.

Another way of looking at this is that one radian is equal to $\frac{180}{\pi}$ degrees = 57.3°.

- To convert from degrees to radians, divide by 360° and multiply by 2π.
- To convert from radians to degrees, divide by 2π and multiply by 360°.

Unlike degrees, where the '°' symbol is universally used, radians are shown in different ways. The name 'radian' or 'rad' may be used, for example, 1.5 rad. Sometimes a superscript lower case 'c' is used, for example, 1.5^c. When an angle is given as a multiple of π, it is understood that this refers to radians and the 'rad' is omitted.

Using radians on your calculator

It makes no difference to most calculations whether angles are measured in degrees or in radians. For example, the sine of an angle is the same whether it is measured in degrees or in radians. The only problem is that when using your calculator, you must ensure that it is set correctly. Most calculators normally work in degrees. This is shown by a small 'D' somewhere on the display as shown circled in Figure 6.3.

Figure 6.3 The display on a calculator will tell you if it is set to work in degrees or in radians. On this calculator pressing 'shift' and 'setup' and then '4' changes it to work in radians

Sometimes it is essential to work in radians, and significant errors arise when this is forgotten. An example of this is when using the expression for simple harmonic motion $y = A\sin(2\pi ft)$ to find the displacement y of an oscillator at time t when the frequency of the oscillator is f and the amplitude is A.

It is essential to realise that part of the expression, $2\pi ft$, is in radians — not in degrees.

For example, for an oscillator of frequency 10 Hz and amplitude 2 m, the displacement at time $t = 0.31$ s is given by: $y = 2\sin(2\pi \times 10 \times 0.31) = 1.18$ m.

A Worked example

A pendulum of length 0.6 m swings through an arc of length 10 cm.

Calculate the angle through which it swings:

i in radians

Use the formula $\theta = \frac{s}{r}$.

angle in radians $= \theta = \frac{s}{r} = \frac{0.1}{0.6} = \frac{1}{6} = 0.1666\ldots = 0.17\,\text{rad}$

ii in degrees

To change radians to degrees, divide your answer to **i** by 2π and multiply by $360°$.

angle in degrees $= \frac{1/6}{2\pi} \times 360 = 9.5°$

TIP

Make sure your measurements of arc and radius are in the same units.

B Guided questions

Copy out the workings and complete the answers on a separate piece of paper.

1. A car travels 100 m along a circular track of radius 250 m.

   Calculate the angle through which the car moves:

   a in radians
   - Use the formula $\theta = \frac{s}{r}$.
   - s is the distance travelled around the circle, 100 m.
   - r is the radius of the circular path.

   $\theta = \frac{s}{r} = \underline{\qquad}$

   b in degrees

   To change radians to degrees, divide your answer to part **a** by 2π and multiply by $360°$.

   angle in degrees $= \underline{\qquad}$

2. A slope is at an angle of $20°$ to the horizontal.

   Calculate this angle in radians.

   To change degrees to radians, use the formula

   $\theta \text{ in radians} = \frac{\text{angle in degrees}}{360°} \times 2\pi$

C Practice questions

3. Convert the following angles in degrees to radians:
   a. 20°
   b. 50°
   c. 90°
   d. 200°
   e. 360°

4. Convert the following angles in radians to degrees:
   a. 0.5 rad
   b. π
   c. 1.2 rad
   d. 4.1 rad
   e. $\frac{3\pi}{2}$

5. A pendulum of length 0.9 m is displaced from the vertical so that the pendulum bob moves through an arc of length 5.0 cm.

   Calculate the angle the string makes with the vertical:
   a. in radians
   b. in degrees

Sine, cosine and tangent

In a right-angled triangle, there are three ratios which are commonly used and are important for A-level Physics. These are sine, cosine and tangent.

The opposite side is always opposite to the angle being measured, the hypotenuse is always the longest side and the adjacent is the side next to the angle that is not the hypotenuse.

$\sin\theta = \dfrac{\text{opposite}}{\text{hypotenuse}}$

$\cos\theta = \dfrac{\text{adjacent}}{\text{hypotenuse}}$

$\tan\theta = \dfrac{\text{opposite}}{\text{adjacent}} = \dfrac{\sin\theta}{\cos\theta}$

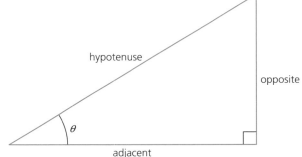

Figure 6.4 Sides of a right-angled triangle for defining sine, cosine and tangent

To find $\sin\theta$ on your calculator, press the sine button and the display will probably show 'sin('. Now enter the angle in degrees followed by ')' and press '='. Your calculator will show the sine of the angle. A similar process will allow you to find the cosine or the tangent.

Note that the maximum value for either $\sin\theta$ or $\cos\theta$ is 1. For $\sin\theta$ this is when $\theta = 90°$ and for $\cos\theta$ this is when $\theta = 0°$, for $0 \leq \theta < 360°$.

6 Geometry and trigonometry

Knowing the definitions of sine, cosine and tangent is particularly important for finding the components of vectors, for example, forces, a topic which is covered later.

Sometimes you will be given a value of $\sin\theta$, $\cos\theta$ or $\tan\theta$ and asked to calculate the angle from the value given. Mathematically this process is written $\sin^{-1}\theta$, $\cos^{-1}\theta$ or $\tan^{-1}\theta$. $\sin^{-1}\theta$ does **not** mean $\frac{1}{\sin\theta}$ which would be written $(\sin\theta)^{-1}$ but has the specific meaning of 'the angle whose sine is θ'.

On your calculator there will be a button for carrying out this process, usually marked '$\sin^{-1}$' but you may have to press 'shift' to access $\sin^{-1}$. You need to ensure that you know which mode your calculator is set to: $\sin^{-1} 0.5$ displays 30 (°) if your calculator is set to degrees but displays 0.52 (rad) or $\frac{1}{6}\pi$ if your calculator is set to radians mode. These do, of course, represent the same angle but you need to be aware of which unit you are using.

Angles greater than 90°

We sometimes need to calculate values of sine, cosine and tangent for angles greater than 90°, and the formulae given work only for acute angles, i.e. $\theta < 90°$. However, if we plot graphs of $\sin\theta$, $\cos\theta$ and $\tan\theta$ for angles between 0° and 360° we obtain the graphs shown in Figure 6.5.

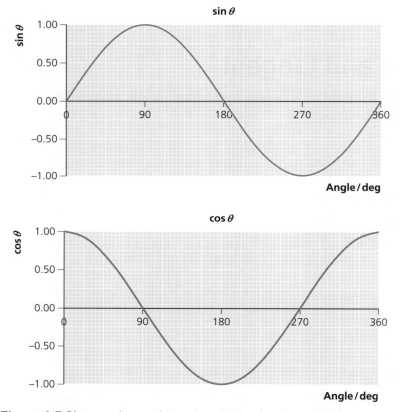

Figure 6.5 Sine, cosine and tangent can be found for angles greater than 90°

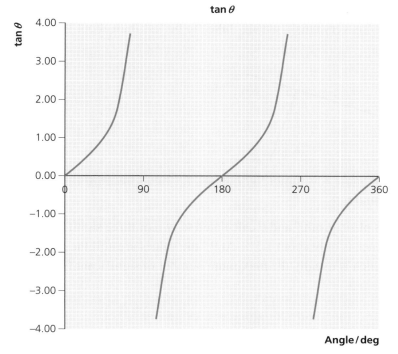

Figure 6.5 (Continued)

The graph of $\sin\theta$ shows that the value of sine is positive for angles between 0° and 180°. It also shows that the value of $\sin\theta$ is the same for two different angles between 0° and 180°. For example, $\sin 30°$ is 0.5 and so is $\sin 150°$. We can see that $150° = 180° - 30°$. The values of $\sin 30°$ and $\sin 150°$ are the same because of the symmetry of the curve around 90°.

$\sin\theta$ is negative for angles between 180° and 360° and, as before, there are two angles for which the sine is the same, for example, $\sin 210°$ is −0.5 and so is $\sin 330°$. So, for example, if you are asked to find $\sin^{-1} 0.707$ there are two answers: 45° and 135°. Normally, unless otherwise indicated, you will be expected to use the angle between 0° and 90° when using sine, cosine or tangent, but you should always bear in mind the other possibilities.

$\cos\theta$ is positive between 0° and 90° and between 270° and 360°. So, for example, you can see from the $\cos\theta$ graph that $\cos 30°$ is 0.87 and $\cos 330°$ is also 0.87.

Again, from the graph, $\cos^{-1}(-0.5)$ is either 120° or 240°.

The graph of $\tan\theta$ has a different shape to sine and cosine. $\tan\theta$ is positive for angles between 0° and 90° and between 180° and 270°. In a similar way to sine and cosine, a value of $\tan\theta$ can refer to two angles between 0° and 360°. For example, $\tan 45°$ is 1 and $\tan 225°$ is also 1.

Small angles

Figure 6.6 shows an angle θ between two lines. The curve, s, is part of the arc of a circle centred at O and with radius $OA = r$, therefore θ in radians is equal to $\frac{\text{arc}}{\text{radius}} = \frac{s}{r}$.

The value of $\sin\theta$ is given by $\frac{\text{opposite}}{\text{hypotenuse}} = \frac{a}{r}$ and $\tan\theta$ by $\frac{\text{opposite}}{\text{adjacent}} = \frac{b}{r}$.

It can be seen from the diagram that $a < s < b$ and therefore $\sin\theta < \theta < \tan\theta$.

However, as we make the angle θ smaller, it can be seen that a, s and b get closer to being the same length and when θ is very small, usually taken to be less than about $10°$, $a \approx b \approx s$ and so $\sin\theta \approx \tan\theta \approx \theta$.

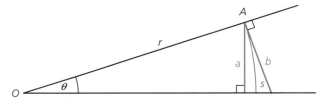

Figure 6.6 Small angles: if angle θ becomes very small, a, b and s become almost the same length and $\sin\theta \approx \tan\theta \approx \theta$

Table 6.1 shows the values of $\sin\theta$, $\tan\theta$ and θ calculated to two or three significant figures. At $10°$ the difference is only about 1.5%. At $9°$ the values are still the same to two significant figures, which is usually the maximum number of significant figures used at A-level.

Table 6.1 The values for $\sin\theta$, $\tan\theta$ and θ for small angles

θ/deg	$\sin\theta$	$\tan\theta$	θ/rad
0	0.000	0.000	0.000
1	0.017	0.017	0.017
2	0.035	0.035	0.035
3	0.052	0.052	0.052
4	0.070	0.070	0.070
5	0.087	0.087	0.087
6	0.105	0.105	0.105
7	0.122	0.123	0.122
8	0.139	0.141	0.140
9	0.156	0.158	0.157
10	0.174	0.176	0.175

This is useful at A-level, for example, when deriving the expression for interference maxima from a diffraction grating, or for the derivation of the expression for the periodic time of a simple pendulum, $T = 2\pi\sqrt{\dfrac{l}{g}}$, which you are likely to encounter in your A-level Physics course. As it will be covered in your physics textbook, it is not repeated here.

A Worked examples

a Find the sine of the angle $35°$.

Make sure your calculator is in degrees mode.

Enter: $\sin(35)$

$\quad = 0.57$

b Find the sine of the angle $290°$.

Enter: $\sin(290)$

$\quad = -0.94$

c **Find the angle whose sine is 0.85.**

Enter: sin⁻¹(0.85)

= 58°

d **Find the sine of the angle 1.2 rad.**

Change your calculator to radians mode.

Enter: sin(1.2)

= 0.93

e i **Find sin 8.5°.**

Change your calculator to degrees mode.

Enter: sin(8.5)

= 0.15

ii **Find tan 8.5°.**

Enter: tan(8.5)

= 0.15

iii **Find sin(0.15 rad).**

Change your calculator to radians mode.

Enter: sin(0.15)

= 0.15

B Guided questions

Copy out the workings and complete the answers on a separate piece of paper.

1 Find the cosine of the angle 40°.

Make sure your calculator is in degrees mode.

Enter: cos(40)

2 Find the cosine of the angle 140°.

Enter: cos(140)

Note that your calculator will correctly give a negative result.

3 In Figure 6.7, side A is 6.0 cm and side B is 7.2 cm.

Calculate cos θ.

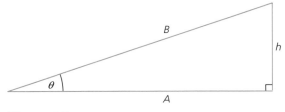

Figure 6.7

Remember that A is the adjacent side and B is the hypotenuse.

Use the formula $\cos\theta = \dfrac{\text{adjacent}}{\text{hypotenuse}}$.

4 **Calculate the angle for which $\sin\theta = 0.866$.**

Make sure your calculator is in degrees mode.

Step 1: enter: $\sin^{-1}(0.866) =$ _____

The display will show the angle in degrees.

Remember that there are two possible angles between $0°$ and $360°$ for which $\sin\theta = 0.866$.

Step 2: check Figure 6.5 to see what the second angle could be.

5 **Refer to Figure 6.7. Side B is 7.2 cm and angle θ is 25°.**

   Calculate h.

Step 1: re-arrange the formula for $\sin\theta$ to give a formula for h.

$\sin\theta = \dfrac{\text{opposite}}{\text{hypotenuse}} = \dfrac{h}{7.2}$ therefore $h = 7.2 \times \sin\theta$

Step 2: calculate $\sin 25°$.

Step 3: use this to find h in cm.

6 **Find the sine of the angle 0.125 rad.**

You can do this in two ways:

- Make sure your calculator is in radians mode.

  Enter: $\sin(0.125)$

- Alternatively you know that, for small angles, θ in radians $\approx \sin\theta$.

  Therefore, without further calculation, $\sin\theta =$ _____.

C Practice questions

7 Refer to Figure 6.7. The distance h is 5.0 cm and the distance B is 1.0 m.
   a Calculate the angle θ:
      i   in degrees
      ii in radians
   b Calculate the distance A.

8 A car going up a hill travels a distance of 800 m. At the same time the vertical height gained is 200 m.

   Calculate the angle of the hill.

9 Figure 6.8 shows two forces, F_1 and F_2, acting at right angles to each other. The resultant of the two forces is R.

   $F_1 = 20\,\text{N}$ and $F_2 = 8.0\,\text{N}$

   a Calculate the angle θ.
   b Calculate the magnitude of R.

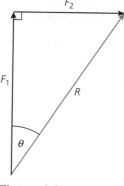

Figure 6.8

10 A distant building is measured to be 100 m away. The 'sighting angle' from the ground to the top of the building is 12°.

Calculate the height of the building.

Pythagoras

A useful property of right-angled triangles allows the length of one side to be calculated if the lengths of the other two sides are known.

Pythagoras's theorem states that $A^2 = B^2 + C^2$ where A is the hypotenuse (see Figure 6.9). This equation can be re-arranged in a number of ways, for example:

$A = \sqrt{B^2 + C^2}$
$B = \sqrt{A^2 - C^2}$
$C = \sqrt{A^2 - B^2}$

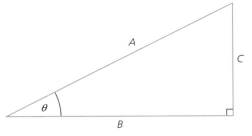

Figure 6.9 A right-angled triangle

An interesting situation occurs if B is 4 units in length and C is 3 units in length. In this case, $A = \sqrt{B^2 + C^2} = \sqrt{4^2 + 3^2} = \sqrt{16 + 9} = \sqrt{25} = 5$

A right-angled triangle with sides in these proportions may be referred to as a '345' triangle.

Note that Pythagoras's theorem is only true for right-angled triangles. In any other triangle, either the sine rule or the cosine rule is required to calculate unknown sides or angles. However, you are not required to use the sine rule or cosine rule in A-level Physics.

A useful result

Based on Figure 6.9 above, $\sin\theta = \frac{C}{A}$ therefore $C = A\sin\theta$.

We also have $\cos\theta = \frac{B}{A}$ and therefore $B = A\cos\theta$.

However, from Pythagoras's theorem, $A^2 = B^2 + C^2$ and so, substituting for B and C, we get $A^2 = A^2\cos^2\theta + A^2\sin^2\theta$.

Dividing both sides by A^2 gives $1 = \cos^2\theta + \sin^2\theta$ or, as it is more usually written, $\sin^2\theta + \cos^2\theta = 1$. This is always true and is a useful identity that you may see used, for example, when adding the kinetic energy and potential energy of an oscillator to give the total energy.

Similar triangles

Similar triangles are useful in A-level Physics. One such use is in determining the formula for linear magnification.

Two triangles are similar if any one of the following three rules is true:
- The three angles of one triangle are the same as the three angles of the other.
- Three pairs of corresponding sides are in the same ratio.
- An angle of one triangle is the same as the angle of the other triangle and the sides containing these angles are in the same ratio.

When triangles are similar, their sides are all in the same ratios. Take, for example, the diagram in Figure 6.10 of a lens magnifying an object. The height of the object is h_O and the height of the image is h_I. The angles marked α are the same in both triangles and this means that all the angles are the same in both triangles and so the triangles are similar according the first rule on the previous page.

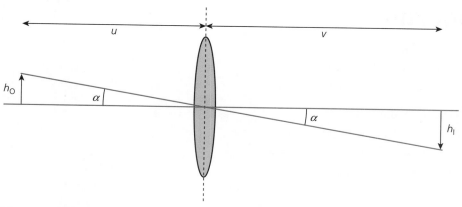

Figure 6.10 Similar triangles

The linear magnification, m, of an object is given by

$$m = \frac{\text{image height}}{\text{object height}}$$

and so by using similar triangles

$$m = \frac{\text{image height}}{\text{object height}} = \frac{h_I}{h_O} = \frac{v}{u}$$

since the ratio of any two corresponding sides in similar triangles must be the same.

A Worked example

In Figure 6.8, assume that $F_1 = 10\,\text{N}$ and $F_2 = 8.0\,\text{N}$. Calculate the magnitude of R.

Use $A = \sqrt{B^2 + C^2}$.

Here R is the hypotenuse (A), and F_1 and F_2 are B and C.

$$R = \sqrt{F_1^2 + F_2^2} = \sqrt{10^2 + 8.0^2} = \sqrt{100 + 64} = \sqrt{164} = 12.8\,\text{N}$$

B Guided questions

Copy out the workings and complete the answers on a separate piece of paper.

1 In Figure 6.9, assume that sides B and C are both $10\,\text{m}$.

   a Calculate the length of side A.

   Use $A = \sqrt{B^2 + C^2}$

   b Calculate the angle between sides A and B.

   Use $\tan\theta = \dfrac{\text{opposite}}{\text{adjacent}} = \dfrac{C}{B}$

   $\theta = \tan^{-1}\left(\dfrac{10}{10}\right) = \underline{\hspace{2cm}}$

c **Calculate the angle between sides *A* and *C*.**

Remember that all the angles in a triangle add up to 180°.

You now have two of the angles since the angle between *B* and *C* is 90°.

Practice questions

2   Refer to Figure 6.9.
    a   If $C = 12$ cm and $B = 16$ cm, calculate A and θ.
    b   If $A = 90$ cm and $C = 15$ cm, calculate B.
    c   If $B = 75$ cm and $A = 95$ cm, calculate C.
    d   If $B = 150$ m and θ is 22°, calculate C and A.
    e   If $B = 36$ m and $A = 45$ m, calculate C.
    f   If $B = 6$ m and $C = 2$ m, calculate A, θ and the angle between A and C.

3   A girl stands 250 m from the bottom of a vertical cliff and measures the angle to the top of the cliff as 8° from the ground.
    a   Calculate the height of the cliff.
    b   Calculate the distance from the top of the cliff to the girl.
    c   A boy stands at the top of the cliff and shouts to the girl. Sound travels at 340 m s^{-1} in air. Calculate the time taken for the shout to reach the girl.

4   The kinetic energy, E_k of an oscillator is $E_k = \frac{1}{2}m\omega^2 A^2 \cos^2 \omega t$.

    The potential energy, E_p of an oscillator is $E_p = \frac{1}{2}m\omega^2 A^2 \sin^2 \omega t$.

    The total energy of an oscillator, $E_T = E_p + E_k$.

    By using $\sin^2 \theta + \cos^2 \theta = 1$, show that $E_T = \frac{1}{2}m\omega^2 A^2$.

5   Refer back to Figure 6.10, in which an object, h_o, of height 2.0 m is 10 m from a lens of a mobile phone camera. In this case the image is formed a distance 1.0 mm behind the lens.

    Calculate the height, h_I, of the image.

Resolving vectors

A vector is a quantity, such as velocity or force, which has both magnitude and direction. For example, a velocity might be 10 m s^{-1} in a direction 60° north from east. This is a vector quantity because it has both magnitude, 10 m s^{-1}, and direction, 60° north from east.

When adding or subtracting vectors, both the direction and magnitude have to be taken into account and the resultant is the sum of the vectors. See Figure 6.8 for an example of adding vectors.

The reverse process is finding two vectors that, when added together, make the original vector. This process is known as resolving vectors and the two vectors which added together make the original vector are called its components. For A-level Physics you will only be expected to resolve co-planar vectors and the required components will be at right angles to each other.

The idea is shown in Figure 6.11.

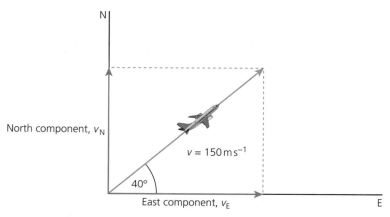

Figure 6.11 Resolving vectors

Imagine a plane travelling at a velocity, v, of $150\,\text{m s}^{-1}$ at an angle of $40°$ north from east. We can separate out its component in the northerly direction, v_N, from its component in the easterly direction, v_E. When added together, the component of the velocity north and the component of the velocity east give the resultant of $150\,\text{m s}^{-1}$ at an angle of $40°$ north from east.

The east component is given by

$$v_E = v \cos 40° = 150 \times \cos 40° = 115\,\text{m s}^{-1}$$

The north component is given by

$$v_N = v \sin 40° = 150 \times \sin 40° = 96\,\text{m s}^{-1}$$

If you use Pythagoras's theorem, you will find that $v = \sqrt{v_N^2 + v_E^2}$.

In general, the component of vector F in the x-direction is given by $F_x = F \cos\theta$ and this is the x or *horizontal* component of F if the x-direction is horizontal.

The component of vector F in the y-direction is given by $F_y = F \sin\theta$ and this is the y or *vertical* component for F if the y-direction is vertical.

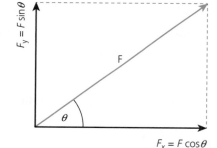

Figure 6.12 Resolving any vector

Although we may want to find the horizontal and vertical components of a vector, this is not always the case.

The x- and y-directions can be at any angle to the horizontal or vertical but must always be at right angles to each other and the vector that is to be resolved is at an angle between the x- and the y-directions. An example is shown in Figure 6.13.

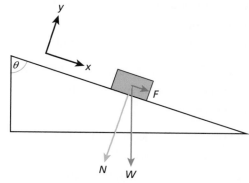

Figure 6.13 Resolving forces for an object on a slope

Figure 6.13 shows a block in equilibrium on a slope. The weight, W, of the block acts directly downwards. There is a component, $F = W_x$, of the weight acting down the slope which must be balanced by a frictional force, $-F$, up the slope if the block is in equilibrium. $F = W_x = W\cos\theta$ since the angle between W and F is also θ. Since W is downwards, $W\cos\theta$ is in the x-direction or down the slope. The choice of making the x-direction down the slope is because we know we want to find the component of the force down the slope and the component at right angles, the y-direction, will be in the direction of the normal to the slope which will be the direction of the contact force.

The y-component of W is $W_y = N$ and, if the block is in equilibrium, this must be balanced by a contact force, $-N$, of the slope on the block. $N = W_y = W\sin\theta$ and, since W is downwards, $W\sin\theta$ is in the $-y$-direction.

A Worked examples

a In Figure 6.14, F is a force of 50 N on a trolley at an angle of 40° to the horizontal.

Figure 6.14 A force acting on a trolley

The trolley is moving at a steady speed, so the horizontal component of force F is equal and opposite to the frictional force.

Calculate the frictional force.

Use the formula $F_x = F\cos\theta$ to calculate the horizontal component.

The horizontal component of the force, F_H, is therefore $F_H = 50 \times \cos 40° = 38$ N.

The frictional force is equal and opposite to this.

b Figure 6.15 shows a weight, W, of 25 N resting on a slope which is at an angle of 30° to the horizontal.

Find the component of the weight perpendicular to the slope, N, and the component down the slope, F.

Step 1: find the angle between W and F.

The angle between W and F is $(90 - 30°) = 60°$.

Step 2: use the formula $F_y = F\sin\theta$ to find the component perpendicular to the slope.

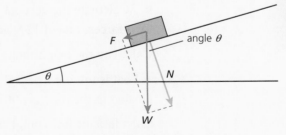

Figure 6.15 A weight on a slope

Component of W perpendicular to the slope
$= W\sin(90° - 30°) = 25 \times \sin 60° = 22$ N

Step 3: use the formula $F_x = F\cos\theta$ to find the component down the slope.

Component of W acting down the slope $= W\cos(90° - 30°) = 25 \times \cos 60° = 12.5$ N ≈ 13 N

B Guided questions

Copy out the workings and complete the answers on a separate piece of paper.

1. **Refer back to Figure 6.15. Assume that W is 500 N and θ is 50°.**

   Calculate the component of the force perpendicular to the slope, N.
   - Use the formula $F_y\, F \sin(90-\theta)$
   - Make sure your calculator is in degrees mode and substitute the given values.

2. Figure 6.16 shows a picture suspended from a string hung around a nail. The weight of the picture is 12 N and the angle the string makes with the horizontal, θ, is 25°.

   a **Show that the tension, T, in the string is about 14 N.**

   Each end of the string supports half the weight therefore the vertical component of T is 6 N.

   $6\,\text{N} = T \times \sin 25°$

   $\Rightarrow T = \dfrac{6}{\sin 25°} = \underline{\qquad}$

   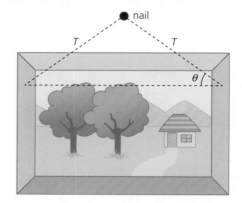

   Figure 6.16 A picture suspended from a nail

   b **Calculate the horizontal component of the tension.**

   If T is 14 N then the horizontal component $F_x = 14 \times \cos 25° = \underline{\qquad}$.

   c **Describe what happens to the tension in the string as the string is shortened.**
   - The vertical component must remain as 6 N but, as the string is shortened, the angle, θ, gets smaller and so $\sin\theta$ gets smaller.
   - As $\sin\theta$ gets smaller, describe what happens to T.

C Practice questions

3. A ball is thrown upwards at a velocity of $8.0\,\text{m s}^{-1}$ at an angle of 40° to the horizontal.
   a Calculate the horizontal component of the velocity.
   b Calculate the vertical component of the velocity.

4. Two forces, one of 12 N and one of 15 N, act on a barge at right angles to each other.

   Calculate the magnitude of the resultant force.

5. A boat is towing a paraglider with a rope at an angle of 70° to the horizontal. The tension in the rope is 600 N.

   Calculate the horizontal component of the force of the rope on the paraglider.

6. A bullet is fired upwards from the ground at 30° to the horizontal at $90\,\text{m s}^{-1}$.

   It hits the ground again 9.2 s later.

   Calculate the distance it has travelled horizontally when it hits the ground.

   $$\text{velocity} = \dfrac{\text{displacement}}{\text{time}}$$

Areas and volumes of simple shapes

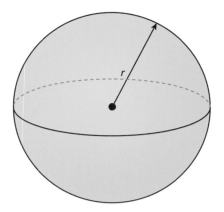

Figure 6.17 A sphere

The area of a circle is πr^2 and the circumference is $2\pi r$ where r is the radius of the circle.

The golden rule is that if there is a (length)^2 in the formula, it must be an area, and if there is a (length)^3 in the formula, it must be a volume. For example, $4\pi r^2$ is the surface area of a sphere, and $\frac{4}{3}\pi r^3$ is the volume of a sphere, where r is the radius of the sphere (see Figure 6.17).

> **TIP**
>
> In the heat of an examination, it's easy to use the wrong formula for area and volume of simple shapes such as a circle or a sphere, so check each formula you use.

Triangles, parallelograms and trapeziums

Calculating the area of a right-angled triangle is straightforward: it is given by $\frac{1}{2}b \times h$ or, of course, $\frac{1}{2}h \times b$, where b is the base of the triangle and h is the height (see Figure 6.18). The reason for this is that a right-angled triangle is half of a rectangle of sides b and h, which would have area $b \times h$.

An isosceles triangle (see Figure 6.19) is made up of two right-angled triangles, so the area is given by $\frac{1}{2}b \times h$.

Suppose, however, you have a triangle that is not a right-angled triangle, such as an obtuse triangle (see Figure 6.20). Here it might not be so obvious that the area of this triangle is also given by $\frac{1}{2}b \times h$ or $\frac{1}{2}h \times b$, where h is the *perpendicular* height of the triangle.

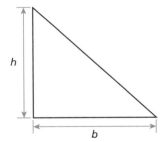

Figure 6.18 A right-angled triangle

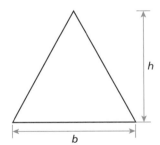

Figure 6.19 An isosceles triangle

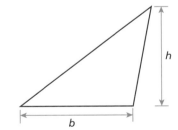

Figure 6.20 An obtuse triangle

A parallelogram has two sets of parallel sides but does not have any right angles (see Figure 6.21). The area of this shape is given by $b \times h$, which you can check by dividing it into two right-angled triangles and a rectangle. You should then be able to see that the obtuse triangle in Figure 6.20 is just half of a parallelogram and therefore half of the area.

A trapezium has one set of parallel sides (see Figure 6.22). It may or may not have a right angle. It is not always easy to calculate the area of a trapezium by dividing it into right-angled triangles and a rectangle without more information about the angles. However, fortunately there is a simple way of finding the area and this is given by $\frac{1}{2}(a+b)h$. Put into words, it is the average of the length of the parallel sides multiplied by the perpendicular distance between the two parallel sides.

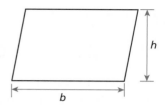

Figure 6.21 A parallelogram

One place where the formula for the area of a trapezium is frequently used is in finding the area under a velocity–time graph (see Figure 6.23) in order to calculate the displacement of a body.

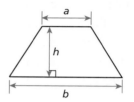

Figure 6.22 A trapezium

Here, the area under the graph is a trapezium, and the area of the trapezium represents the displacement, s.

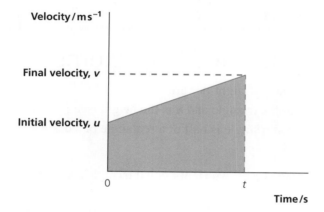

Figure 6.23 The area under a velocity–time graph

The parallel sides have length v and length u, so the average length is $\frac{v+u}{2}$.

The area of the trapezium is therefore given by $s = \frac{v+u}{2} \times t$ and this is one of the familiar equations of motion.

Cylinders

A uniform wire is one example of a cylinder (see Figure 6.24). The cross-sectional area may be required, for example, to calculate the tensile stress. The volume of the wire may also be needed. For example, the area under a stress–strain graph is the work done per unit volume in stretching a wire. This will be covered in detail in your A-level textbook. Here we concentrate on how to calculate volumes.

The cross-sectional area is the area of the end of the cylinder, shaded in Figure 6.24, and is given by πr^2 where r is the radius of the wire. If you have measured the diameter, d, of the wire, the cross-sectional area is given by $\pi \frac{d^2}{4}$.

The volume of a cylinder is given by $\pi r^2 h$.

The total surface area of a cylinder is given by $2\pi r(r + h)$.

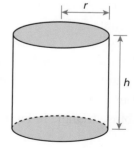

Figure 6.24 A cylinder

Full worked solutions at www.hoddereducation.co.uk/essentialmathsanswers

A Worked examples

a The diameter of a wire is 3.2 mm.

Calculate the cross-sectional area of the wire.

Step 1: convert units to base units where necessary.

Convert 3.2 mm into metres.

$$3.2 \text{ mm} = 3.2 \times 10^{-3} \text{ m}$$

Step 2: calculate the radius.

$$r = \frac{d}{2} = \frac{3.2 \times 10^{-3}}{2} = 1.6 \times 10^{-3} \text{ m}$$

Step 3: use the formula $A = \pi r^2$ for the cross-sectional area of a cylinder.

$$A = \pi \times r^2 = \pi(1.6 \times 10^{-3})^2 = 8.0 \times 10^{-6} \text{ m}^2$$

b In a trapezium like the one shown in Figure 6.22, $a = 3$ cm, $h = 3$ cm and $b = 5$ cm.

Calculate the area of the trapezium.

Step 1: use the formula $\frac{1}{2}(a + b)h$ for the area of a trapezium.

Step 2: substitute the numbers and calculate the area.

$$A = \frac{1}{2}(3 + 5) \times 3 = 12 \text{ cm}^2$$

c A ball bearing has a diameter of 0.5 cm.

Calculate the volume of the ball bearing.

Step 1: convert units to base units where necessary.

Convert 0.5 cm to metres.

$$0.5 \text{ cm} = 0.5 \times 10^{-2} \text{ m}$$

Step 2: calculate the radius.

$$r = \frac{d}{2} = \frac{0.5 \times 10^{-2}}{2} = 2.5 \times 10^{-3} \text{ m}$$

Step 3: use the formula $V = \frac{4}{3}\pi r^3$ to calculate the volume of a sphere.

$$V = \frac{4}{3}\pi \times (2.5 \times 10^{-3})^3 = 6.5 \times 10^{-5} \text{ m}^3$$

B Guided questions

Copy out the workings and complete the answers on a separate piece of paper.

1 Calculate the cross-sectional area of a wire of diameter 2.4 mm.

Step 1: convert units to base units where necessary.

2.4 mm = _____ m

Step 2: calculate the radius, r, of the wire.

$$r = \frac{d}{2} = \text{_____}$$

Step 3: use the formula for the cross-sectional area of a cylinder.

$$A = \pi r^2 = \text{_____}$$

2 Calculate the volume of a ball bearing of diameter 2.5 mm.

Step 1: convert units to base units where necessary.

2.5 mm = _____ m

Step 2: calculate the radius, r, for the ball bearing.

$r = \dfrac{d}{2} =$ _____

Step 3: use the formula for the volume of a sphere.

$V = \dfrac{4}{3}\pi r^3 =$ _____

3 The trapezium in Figure 6.22 has sides $a = 5.0$ cm, $b = 8.0$ cm and $h = 7.0$ cm.

Calculate the area of the trapezium in cm².

Use the formula for the area of a trapezium.

$A = \dfrac{1}{2}(a+b)h =$ _____

4 A wire of length 2.5 m has a diameter of 3.5 mm.

Calculate the volume of the wire.

Step 1: calculate the radius, r, of the wire.

$r = \dfrac{d}{2} =$ _____

Step 2: use the formula for the volume of a cylinder.

$V = \pi r^2 h$, where $h = 2.5$ m

$V =$ _____

C Practice questions

5 A car increases its speed uniformly from 8.0 m s⁻¹ to 14 m s⁻¹ in 15 s.

Calculate the displacement of the car over the 15 s.

6 The resistance, R, of a wire is given by $R = \dfrac{\rho l}{A}$, where ρ is the resistivity, l is the length and A is the cross-sectional area.

Calculate the resistance of a piece of copper wire of length 0.5 m and diameter 0.5 mm. The resistivity of copper is $1.7 \times 10^{-8}\,\Omega\,\text{m}$.

7 The mean radius of the Earth is 6371 km.

Calculate the volume of the Earth in m³.

8 A piece of metal has the shape shown in Figure 6.20.

The distance b is 0.50 m and the distance h is 0.4 m.

Calculate the area of the piece of metal.

9 A rectangular block has length 5.0 cm, width 4.0 cm and height 3.0 cm.
   a   Calculate the surface area of the block.
   b   Calculate the volume of the block.

7 Exponential changes

Note: this Unit is assessed at A-level only.

When the rate of change of a quantity is directly proportional to the amount of the quantity present, this is a process of exponential change.

Radioactive decay is one example of an exponential change but you will encounter others. For example, when a capacitor is discharging, the rate of change of charge on a capacitor is proportional to the charge on the capacitor. Exponential changes do not have to be a decay: population growth, such as the unrestricted growth of bacteria, is an example of exponential growth. For either growth or decay, the mathematical analysis is similar. We will look first at radioactive decay.

Radioactive decay

There are two assumptions for considering radioactive decay:
- The rate of decay $\frac{\Delta N}{\Delta t}$ is proportional to the number, N, of unstable nuclei present.
- Radioactive decay is a random process governed by probability.

Taking the first assumption, we can say that $\frac{\Delta N}{\Delta t} \propto -N$, where the negative sign indicates that N is decreasing with time.

The constant of proportionality is λ and is called the decay constant, which is a different constant for each radioactive isotope. Therefore

$$\frac{\Delta N}{\Delta t} = -\lambda N$$

If we make the time interval, Δt, smaller and smaller ($\Delta t \to 0$), this equation becomes $\frac{dN}{dt} = -\lambda N$ which is known as a first-order differential equation. As explained in Unit 5 Graphs, 'Rates of change', $\frac{dN}{dt}$ is the rate of change of N with time. You do not need to know how to solve this equation but you do need to use both it and its solution.

The solution to the equation is

$$N = N_0 e^{-\lambda t}$$

where N is the number of nuclei left to decay at time t and N_0 is the initial number of nuclei at time $t = 0$. The symbol e in the equation is an irrational number which is approximately 2.718 and is sometimes called the 'natural number' because it occurs in the mathematical solution to any process where the rate of change depends on the amount present — as happens in many natural processes.

Although $N = N_0 e^{-\lambda t}$ is called the 'solution' to $\frac{dN}{dt} = -\lambda N$, do not let that confuse you. Both equations mathematically describe exactly the same change or, if you prefer, refer to exactly the same line on a graph of N against t. A graph showing the shape of this line is shown in Figure 7.1.

Activity

The activity, A, of a source is defined as the number of disintegrations per second and is measured in becquerels (Bq), where one Bq is one decay per second. It is a positive number and so $A = -\frac{\Delta N}{\Delta t}$, which means that

$$A = \lambda N$$

A graph of activity against time, which is what is usually measured, will have the same shape as the graph in Figure 7.1. Since $A \propto N$ we can also say that

$$A = A_0 e^{-\lambda t}$$

where A is the activity at time t and A_0 is the initial activity at time $t = 0$.

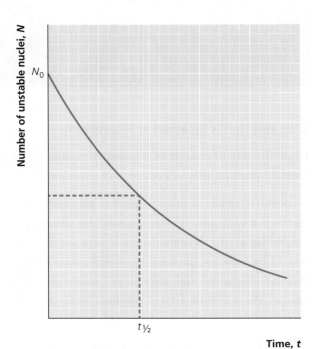

Figure 7.1 The decay curve for a radioactive material

Half-life

The equation $N = N_0 e^{-\lambda t}$ can be re-arranged to give $\frac{N}{N_0} = e^{-\lambda t}$. The first thing to notice about this is that, for a given value of t, $e^{-\lambda t}$ is always the same for a particular radioactive isotope, and so $\frac{N}{N_0}$ is always the same. It doesn't matter when we start timing, if we measure the number of nuclei present at $t = 0$ and then the number present after time t, the ratio $\frac{N}{N_0}$ will always be the same for a particular radioactive isotope.

This is a feature of exponential decay; equal time intervals will always result in equal fractional or percentage changes in N. This leads to the idea of half-life. We define the half-life ($t_{\frac{1}{2}}$) as the time taken for the number of unstable nuclei to fall to half or 50% of N_0. This is shown by the dashed line in Figure 7.1. If the value of N is $\frac{N_0}{2}$ after a certain time, $t_{\frac{1}{2}}$, then

$$\frac{N_0}{2} = N_0 e^{-\lambda t_{\frac{1}{2}}}$$

which becomes

$$\frac{1}{2} = e^{-\lambda t_{\frac{1}{2}}}$$

if we divide both sides of the equation by N_0. But we know that $e^{-\lambda t} = \frac{1}{e^{\lambda t}}$ and so we can find the reciprocal of both sides of the equation to give

$$2 = e^{\lambda t_{\frac{1}{2}}}.$$

In Unit 8 Logarithms, you will learn that the natural log (ln) of $e^{\lambda t_{\frac{1}{2}}}$ is $\lambda t_{\frac{1}{2}}$ so, taking logs to the base e of both sides, we get

$$\ln 2 = \lambda t_{\frac{1}{2}} \qquad \text{or} \qquad t_{\frac{1}{2}} = \frac{\ln 2}{\lambda}$$

This can be further re-arranged to give $\lambda = \frac{\ln 2}{t_{1/2}}$.

A Worked examples

a **1 g of technicium-99 contains 6.1×10^{21} atoms. The decay constant, λ, is $3.2 \times 10^{-5}\,\text{s}^{-1}$.**

Calculate the initial activity of the sample.

Use $A = \lambda N$ and substitute the given values.

$A = 3.2 \times 10^{-5} \times 6.1 \times 10^{21} = 2.0 \times 10^{17}\,\text{Bq}$

b **A sample of radioactive material has 5.0×10^{20} nuclei. The decay constant, λ, is $0.069\,\text{s}^{-1}$.**

i **Calculate the half-life of the material.**

- Use $t_{\frac{1}{2}} = \frac{\ln 2}{\lambda}$ and substitute the given values.
- The value of $\ln 2$ is 0.693.

$t_{\frac{1}{2}} = \frac{0.693}{0.069} = 10\,\text{s}$

ii **Calculate the number of unstable nuclei left after 15 s.**

- Use $N = N_0 e^{-\lambda t}$ and substitute the given values.
- Using the given values for N_0, λ and t, we have $N = 5.0 \times 10^{20} \times e^{-0.069 \times 15}$.

Step 1: to evaluate $e^{-0.069 \times 15}$, press the 'e^x' button on your calculator, then enter '-0.069×15' and press the equals button.

Step 2: multiply the answer by 5.0×10^{20} to give $N = 1.8 \times 10^{20}$ nuclei.

c **The decay constant of the radioactive isotope cobalt-60 is $4.2 \times 10^{-9}\,\text{s}^{-1}$.**

A newly prepared 5 g sample of cobalt-60 contains 5.0×10^{22} nuclei.

i **Calculate the number of disintegrations per second initially measured from the sample.**

- The number of disintegrations in the first second is the activity.
- Use $A = \lambda N$ and substitute the given values.

$A = \lambda N = 4.2 \times 10^{-9} \times 5.0 \times 10^{22} = 21 \times 10^{14}\,\text{s}^{-1}$

The number of disintegrations per second is $2.1 \times 10^{14}\,\text{Bq}$.

ii **Explain why the number of disintegrations per second decreases with time.**

The number will decrease because, although the decay constant is unchanged, the number left to decay decreases and therefore the number that decay must also decrease.

B Guided questions

Copy out the workings and complete the answers on a separate piece of paper.

1 Strontium-90 is a source of beta radiation and has a half-life of 29 years.

A school source has an initial activity of $1.9 \times 10^5\,\text{Bq}$.

a Calculate the decay constant for strontium-90.

Step 1: convert 29 years to seconds.

$29 \times 365 \times 24 \times 3600 = 9.1 \times 10^8\,\text{s}$

Step 2: use $\lambda = \frac{\ln 2}{t_{1/2}}$ and substitute the value for the half-life.

Remember $\ln 2 = 0.693$.

b Calculate the activity after five years.

Step 1: convert five years to seconds.

$5 \times 365 \times 24 \times 3600 = 1.6 \times 10^8$ s

Step 2: use $A = A_0 e^{-\lambda t}$ and substitute the given values.

$A = 1.9 \times 10^5 \times e^{-7.6 \times 10^{-10} \times 1.6 \times 10^8} =$ _____

See worked example **bii** for how to evaluate e^x. Remember the unit for activity.

2 Iodine-131 is used for the treatment of thyroid cancer. It has a half-life of eight days. A 0.5 mg sample of iodine-131 contains 2.3×10^{18} unstable nuclei.

a Calculate the initial activity of the sample.

Step 1: convert eight days to seconds.

$8 \times 24 \times 3600 =$ _____ s

Step 2: use $\lambda = \dfrac{\ln 2}{t_{1/2}}$ and substitute your value for the half-life.

Remember $\ln 2 = 0.693$.

Step 3: use $A = \lambda N$ and substitute your value for λ and the given value of N. Remember the unit for activity.

b Calculate the number of unstable nuclei left after 24 days.

Step 1: work out the number of half-lives.

24 days is three half-lives.

Step 2: work out the number of unstable nuclei after each half-life.

After one half-life the number of unstable nuclei = $\dfrac{2.3 \times 10^{18}}{2}$

After two half-lives the number of unstable nuclei = $\dfrac{2.3 \times 10^{18}}{2 \times 2}$

After three half-lives the number of unstable nuclei = $\dfrac{2.3 \times 10^{18}}{2 \times 2 \times 2}$

3 A radioactive source initially has 2.4×10^{20} unstable nuclei. After 50 s the number of unstable nuclei is 0.8×10^{20}.

Calculate the decay constant and the half-life for the source.

Step 1: use $N = N_0 e^{-\lambda t}$ and substitute the given values.

Substituting the given values for N, N_0 and t gives $0.8 \times 10^{20} = 2.4 \times 10^{20} \times e^{-\lambda \times 50}$

Step 2: re-arrange to find the value of λ.

- Divide both sides by 2.4×10^{20} to give $= \dfrac{0.8 \times 10^{20}}{2.4 \times 10^{20}} = \dfrac{1}{3} = e^{-\lambda \times 50}$
- Find the reciprocal of both sides.

  $3 = e^{-\lambda \times 50}$

- Take natural logs of both sides (see Unit 8).

  $\ln 3 = \lambda \times 50$

- Divide both sides by 50 to give $\dfrac{\ln 3}{50} = \lambda$.

Step 3: use $t_{\frac{1}{2}} = \dfrac{\ln 2}{\lambda}$ and substitute the value for λ to find the half-life.

Practice questions

4 The activity of a source is initially measured to be 6×10^5 Bq. It falls to 2.5×10^5 Bq in 50 min.
   a Calculate the decay constant for the source.
   b Calculate the activity of the source after two hours.

5 Carbon-14 decays with a half-life of 5730 years. It is used to carbon-date organic material.

   The percentage of carbon-14 in living material is thought to be constant over time but, once the material is dead, the carbon-14 decays and is not replaced.
   a Calculate the decay constant.
   b A 1 g sample of recent organic material is found to have an activity of 0.192 Bq. A 1 g sample of ancient carbon material from an archaeological dig is found to have an activity of 0.184 Bq.

   Calculate the age of the ancient carbon material.

Capacitor discharge

The expression for capacitor discharge is similar to that for radioactive decay. When a charged capacitor is discharged through a resistor, the current through the resistor is given by $I = \frac{V}{R}$, where V is the potential difference across the capacitor and resistor (volts), R is the resistance (ohms) and I is the current (amps). Current is the rate of change of charge, $I = \frac{\Delta Q}{\Delta t}$, where Q is the charge (coulombs) and t is time (seconds). The potential difference across a capacitor is given by $V = \frac{Q}{C}$, where C is the capacitance (farads). Combining these equations, we get

$$\frac{\Delta Q}{\Delta t} = -\frac{Q}{RC}$$

where, as before, the minus sign indicates that Q is decreasing. You should recognise this as being both similar to the equation for radioactive decay and satisfying the condition for an exponential change where the rate of change of a quantity is directly proportional to the quantity present.

The solution to this equation is also similar:

$$Q = Q_0 e^{\frac{-t}{RC}}$$

where Q_0 is the charge on the capacitor at time $t = 0$ and Q is the charge on the capacitor at time t.

Since $V \propto Q$ we can also write

$$\frac{\Delta V}{\Delta t} = -\frac{V}{RC} \quad \text{and} \quad V = V_0 e^{\frac{-t}{RC}}$$

A graph of either Q against t or V against t has the same shape as the graph in Figure 7.1.

RC is known as the time constant. RC must have the units of time because the index for e cannot have a unit and so the unit of RC must cancel out with the unit for t.

If we measure the charge at a time $t = RC$, then $Q = Q_0 e^{\frac{-RC}{RC}} = Q_0 e^{-1} = Q_0 \times 0.37$.

This means that after a time equal to the time constant, the charge or potential difference will have fallen to 0.37 or 37% of its starting value.

7 Exponential changes

A Worked example

A capacitor of value 100 μF, initially charged with 10 μC, is discharged through a resistor of value 50 kΩ.

a Calculate the time constant for the circuit.

Step 1: convert both C and R to base units.

$C = 100 \times 10^{-6}$ F $R = 50 \times 10^{3}$ Ω

Step 2: use time constant = RC.

time constant = $50 \times 10^{3} \times 100 \times 10^{-6}$

$= 5$ s

b Calculate the charge on the capacitor after 8 s.

Use $Q = Q_0 e^{\frac{-t}{RC}}$ and substitute the values for Q_0, t and RC.

$Q = 10 \times 10^{-6} \times e^{\frac{-8}{5}} = 2.0 \times 10^{-6}$ C $= 2.0$ μC

B Guided question

Copy out the workings and complete the answers on a separate piece of paper.

1 A capacitor of value 4700 μF is initially charged to a potential difference of 10 V. It is discharged through a resistor of value 100 kΩ.

a Calculate the time constant for the circuit.

Step 1: convert both C and R to base units.

$C = 4700 \times 10^{-6}$ F $R = 100 \times 10^{3}$ Ω

Step 2: use time constant = RC.

time constant = $100 \times 10^{3} \times 4700 \times 10^{-6} =$ _____

b Calculate the time taken for the potential difference to fall to 5.0 V.

Use $V = V_0 e^{\frac{-t}{RC}}$ and substitute the given values.

- Substitute the values for V, V_0 and RC: $5 = 10 \times e^{\frac{-t}{470}}$
- Divide both sides by 10 to give $\frac{5}{10} = \frac{1}{2} = e^{\frac{-t}{470}}$
- Take the reciprocal of both sides to give $2 = e^{\frac{-t}{470}}$
- Take natural logs of both sides to give $\ln 2 = \frac{t}{470}$
- Re-arrange to find t: _____

C Practice questions

2 Calculate the time constant for the following:
   a A capacitor of value 500 μF discharging through a 10 kΩ resistor.
   b A capacitor of value 50 mF discharging through a 500 Ω resistor.
   c A 1 F capacitor discharging through a 200 Ω resistor.

Full worked solutions at www.hoddereducation.co.uk/essentialmathsanswers

3 A capacitor is initially charged to 15 V. It is discharged through a resistor and after 10 s is found to have a potential difference of 9.0 V across it.
   a Calculate the time constant for the circuit.
   b If the resistance, R, is known to be 120 kΩ, calculate the value of the capacitor, C.

4 The equation for the current, I, at time t in a capacitor charging or discharging circuit is $I = I_0 e^{\frac{-t}{RC}}$, where C is the capacitance and R is the value of the resistor in the circuit. I_0 is the current at time $t = 0$.

A circuit consists of a capacitor of 150 µF and a resistor of 100 kΩ. The initial current is 10 mA.

Calculate the current after 20 s.

8 Logarithms

Note: this Unit is assessed at A-level only.

Understanding logarithms

If you look in any A-level Physics textbook you will find logarithms used in a number of ways. The idea of a logarithm is really quite simple and using logarithms is both useful and sometimes essential. The invention of logarithms is generally attributed to John Napier (1550–1617) who lived in Merchiston Castle in Edinburgh, UK, and who first published his ideas in 1614. In the days before calculators and computers, logarithms made complicated calculations easier, especially calculations for astronomy.

A logarithm (to the base 10) of a specific number (x) is the number (y) to which 10 must be raised in order to get the specific number, x. Mathematically, if $x = 10^y$ then $\log_{10} x = y$, so finding $\log x$ and 10^x are exactly opposite mathematical operations.

For example, $1000 = 10^3$ therefore $\log_{10}(1000) = 3$. Similarly, $\log_{10}(100) = 2$.

The reason that this is helpful in difficult calculations is because by using logs, multiplication and division of large numbers becomes a simple addition or subtraction sum. However, this needs further explanation and you may need to refer back to Unit 1 to remind yourself about powers of 10 and using indices.

Recall that, for example, in the calculation $10^3 \times 10^2 = 10^5$ we add the indices, $3 + 2 = 5$. Similarly, when dividing the two numbers, $\frac{10^3}{10^2} = 10^1$, we subtract the indices in order to carry out the division calculation.

For simple powers of 10 this is straightforward but suppose we wanted to multiply 4640×2320. One way of carrying out the calculation would be to use logs.

To find $\log_{10}(4640)$ on a calculator use the button 'log'. Depending on your calculator, if you press 'log' the display may show 'log('. Enter '4640)', then press the equals button. (On some calculators you may enter the number first and then press 'log'.)

$\log_{10}(4640) = 3.667$ and $\log_{10}(2320) = 3.365$

Adding the logs: $3.667 + 3.365 = 7.032$.

Therefore the result of the original calculation 4640×2320 is $10^{7.032}$.

To find $10^{7.032}$ on your calculator, press 'shift+log' ($=10^x$) and your display will show '10^x'. Now enter '7.032' and press the equals button. (On some calculators you may enter 7.032 first then press shift+log.)

$10^{7.032} = 10\,764\,652$ or 1.076×10^7

Doing the same calculation in the ordinary way on a calculator gives $10\,764\,800$ or 1.076×10^7 to four significant figures. Any small difference is simply because of limiting the number of significant figures in the calculations and, if you keep all the significant figures throughout the calculation, the results will be identical.

Suppose instead we wanted to divide 4640 by 2320. This time we subtract the logs: $3.667 - 3.365 = 0.302$. Therefore the result of this calculation is $10^{0.302} = 2.004472 = 2.00$ (to three significant figures). Doing the same calculation on a calculator also gives 2.00.

Before calculators or computers, there were tables of logarithms published so that students could look up the numbers in a table to find the logarithm. A slide rule, which had a logarithmic scale, was also used to find the results of calculations. Now that we have calculators, we do not use logarithms to carry out everyday calculations but logarithms are still useful and, as we shall see later, understanding how logarithms are used to multiply and divide numbers is still important.

Logarithmic scales

One reason that logarithms are useful is that we can use logarithmic scales for graphs or charts where the range of the graph would be too large if linear.

For example, if we plot the resistivities of common materials on a linear scale, the resulting graph would not be very helpful because the scale is too small to see many of the values (see Figure 8.1).

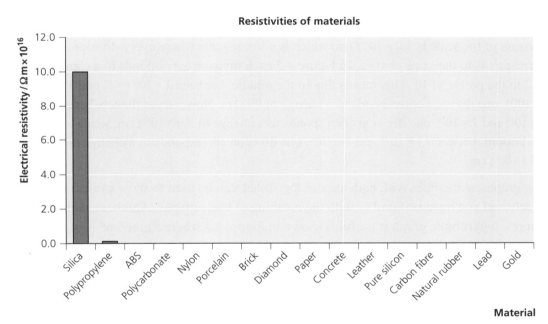

Figure 8.1 The range of resistivity is so great that a linear scale cannot show all the materials on the same chart

But if we plot exactly the same data on a chart with a logarithmic scale, where equal divisions are equal powers of 10, the chart immediately becomes a useful comparison of resistivities.

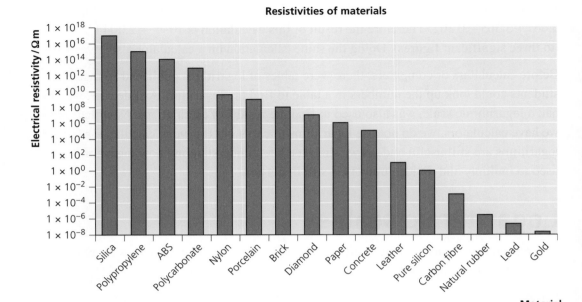

Figure 8.2 With a logarithmic scale, the range of resistivities of the materials can be clearly seen

Look carefully at the scales on the y-axes of Figures 8.1 and 8.2. In Figure 8.1 each division of the scale is $2.0 \times 10^{16}\,\Omega\,\text{m}$, which is a linear scale where every division corresponds to the same change. In Figure 8.2 each division corresponds to a change of 2 in the power of 10. This means that in the middle, between 1×10^0 (= 1) and 1×10^2 (= 100), one division corresponds to a change of $99\,\Omega\,\text{m}$. However, at the top, between 1×10^{16} and 1×10^{18}, one division corresponds to a change of $9.9 \times 10^{17}\,\Omega\,\text{m}$, while at the bottom, between 1×10^{-8} and 1×10^{-6}, one division corresponds to a change of $9.9 \times 10^{-7}\,\Omega\,\text{m}$.

Log graphs, sometimes with both axes as log scales, can be used to show values that vary widely and you must be familiar with the meaning of these graphs. Another example of where a logarithmic graph is useful is shown in Figure 8.3 where a graph of $\frac{1}{f}$ is plotted against wavelength λ for the electromagnetic spectrum.

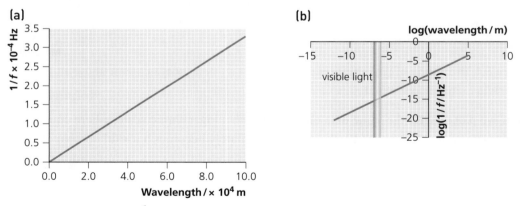

Figure 8.3 A graph of $\frac{1}{f}$ against λ for the electromagnetic spectrum. In (a) the region that corresponds to visible light is too small and too close to the origin to be seen. In (b) the same data is plotted with log graphs, showing the region corresponding to visible light

Full worked solutions at www.hoddereducation.co.uk/essentialmathsanswers

In Figure 8.3a the region that corresponds to visible light cannot be shown because, with a wavelength of around 0.006×10^{-4} m, it is too close to the origin if the whole range of wavelengths is to be shown. However, by plotting a graph of $\log\left(\frac{1}{f}\right)$ against $\log \lambda$, as in Figure 8.3b, the region of visible light can be seen relative to the rest of the spectrum.

A Worked examples

a Calculate the following:

i $10^{3.2}$

Step 1: on your calculator press 10^x and the display will show 10^x.

Step 2: enter '3.2 ='

The display will show $1584.9 \approx 1.6 \times 10^3$.

ii $10^{0.2}$

Step 1: on your calculator press 10^x and the display will show 10^x.

Step 2: enter '0.2 ='

The display will show $1.5849 \approx 1.6$.

iii $10^{-4.2}$

Step 1: on your calculator press 10^x and the display will show 10^x.

Step 2: enter '−4.2 ='

The display will show 6.31×10^{-5}.

b Use logs to find the value of the following:

i 54×80

Step 1: use your calculator to find the log of each number.

$\log(54) = 1.73239$ and $\log(80) = 1.90309$

Step 2: add the two logs to give a value, x.

$x = 1.73239 + 1.90309 = 3.63548$

Step 3: use your calculator to find 10^x.

- On your calculator press 10^x and the display will show 10^x.
- Enter '3.63548 ='
- The display will show 4319.9627.
- The answer is 4320 (to four significant figures).

ii $2385 \div 563$

Step 1: use your calculator to find the log of each number.

$\log(2385) = 3.37749$ and $\log(563) = 2.75051$

Step 2: subtract the logs to give a value, x.

$x = 3.37749 − 2.75051 = 0.62698$

Step 3: use your calculator to find 10^x.

- On your calculator press 10^x and the display will show 10^x.
- Enter '0.62698 ='
- The answer is 4.24 (to three significant figures).

B Guided questions

(Although you won't have to do calculations like question 1 in your A-level Physics, the practice will help you to understand how logs work and help you use logs more confidently.)

Copy out the workings and complete the answers on a separate piece of paper.

1 Use logs to do the following calculations.

   Check your answer by doing the calculation in the ordinary way on your calculator.

   a 120×5

   Step 1: use your calculator to find the log of each number.

   log(120) = _____

   log(5) = _____

   Step 2: add the logs together to give a value, x.

   Step 3: use your calculator to find 10^x. See the text for instructions on how to do this.

   b 3600×24

   Step 1: use your calculator to find the log of each number.

   log(3600) = _____

   log(24) = _____

   Step 2: add the logs together to give a value, x.

   Step 3: use your calculator to find 10^x.

   c $3.2 \times 10^7 \div 365$

   Step 1: use your calculator to find the log of each number.

   log(3.2×10^7) = _____

   log(365) = _____

   Step 2: subtract log(365) from log(3.2×10^7) to give a value, x.

   Step 3: use your calculator to find 10^x.

2 **a Explain how you can tell that the y-axis of the graph in Figure 8.2 is logarithmic.**

   To answer this question, look at the scale markings of the y-axis and note the way that the numbers increase.

   b Use the graph of Figure 8.3b to calculate the wavelength of a wave of frequency 1.0×10^{10} Hz.

   Step 1: calculate $\frac{1}{f}$.

   Step 2: calculate $\log\left(\frac{1}{f}\right)$.

   Step 3: find the value of $x = \log \lambda$ that corresponds to this value of $\log\left(\frac{1}{f}\right)$.

   Step 4: use 10^x to find λ.

C Practice questions

3 Calculate the logs to the base 10 of the following:
   a 0.10
   b 3.00×10^{24}
   c $\dfrac{1}{1.20 \times 10^{-3}}$

4 Use logs to calculate the following:
   a $4.5 \times 10^3 \times 6.02 \times 10^{23}$
   b $76 \times 1.6 \times 10^{-19}$
   c $\dfrac{9.0 \times 10^9 \times (1.6 \times 10^{-19})}{1.0 \times 10^{-9}}$

Using logarithms

The use of log graphs to find or confirm relationships is a useful technique for A-level Physics.

Suppose we speculate that the period, T, of a pendulum might depend on the length, l, of the pendulum. From the results of an experiment, we might plot a graph of T against l (see Figure 8.4). The resulting graph is not a straight line and so little information can be gained from it other than the fact that there is not a linear relationship between T and l.

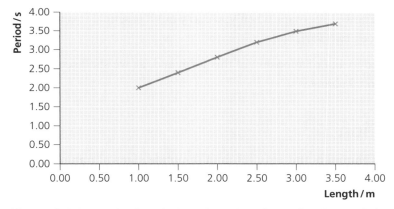

Figure 8.4 A graph of period against time for a simple pendulum

However, if we guess that the periodic time, T, for a pendulum is proportional to l^n, then we can write $T = kl^n$, where k is a constant and n could be any number.

We can take logarithms of the equation $T = kl^n$. Remember that multiplying two numbers is the same as adding their logarithms, so $\log l^n$ is the same as $n \log l$ since l^n is just l multiplied by itself n times and kl^n is the same as $k \times l^n$. Therefore

$\log T = \log k + n \log l$

This equation now looks rather like the standard equation of a straight line, $y = mx + c$, where m is the gradient and c is the intercept (see Unit 5). For our equation, n is the gradient and $\log k$ is the intercept.

If our guess is correct and we now plot $\log T$ against $\log l$, we should get a straight line with gradient n and intercept $\log k$. This is shown in Figure 8.5.

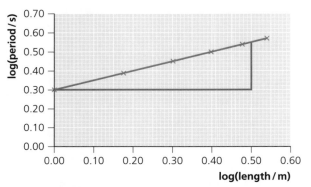

Figure 8.5 A graph of log T against log l

The gradient of this graph, shown by using the green lines, is

$$\frac{\Delta y}{\Delta x} = \frac{(0.55 - 0.30)}{0.5} = 0.5$$

Therefore n is 0.5 or, $T = kl^{0.5} = k\sqrt{l}$.

The intercept is $\log k = 0.3$. Therefore $k = 10^{0.3} = 2.0$.

This gives the final expression as $T = 2.0 \times \sqrt{l}$.

As it happens, we know from the theory of the pendulum that

$$k = \frac{2\pi}{\sqrt{g}} = \frac{2 \times \pi}{\sqrt{9.8}} = 2.0$$

on Earth, so the method has been effective in finding the relationship. In this case, we knew the relationship before we started. However, this is a powerful method to find a relationship that you do not know — perhaps when you carry out a practical investigation to find the relationship between two variables.

This technique can be applied to any relationship of this form and, as we shall see later, is especially useful for exponential changes.

B Guided question

Copy out the workings and complete the answers on a separate piece of paper.

1 The graph in Figure 8.6 is for an experiment in which the spring constant, k, for a mass hanging on a spring is changed and the time period, T, of oscillation measured.

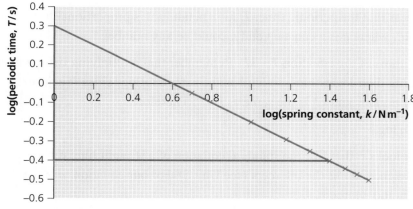

Figure 8.6

Assume that $T = ck^n$ where c is a constant.

a Use the graph to find a value for n.

Step 1: take logs of the equation $T = ck^n$.

You will then have an equation that resembles $y = mx + c$, with n as the gradient and $\log c$ as the intercept.

Step 2: work out the gradient $\frac{\Delta y}{\Delta x}$ for the graph to find n.

b Use the graph to find a value for c.

The intercept on the y-axis is $\log c$.

Therefore $c = 10^{\text{intercept value}}$.

c Write down the complete relationship between T and k.

Substitute your values of c and n in the expression $T = ck^n$.

C Practice question

2 An experiment is carried out to measure the periodic time of the oscillations of a mass on a spring. The mass is changed and the time period is measured.

The results in Table 8.1 are obtained.

Table 8.1

m/kg	T/s	$\log(m/\text{kg})$	$\log(T/\text{s})$
0.1	0.63		
0.3	1.09		
0.5	1.40		
0.7	1.66		
0.9	1.88		
1.1	2.08		
1.3	2.27		
1.5	2.43		

It is suggested that the relationship between T and m is of the form $T = cm^n$, where c is a constant.

a Complete the table for $\log(m/\text{kg})$ and $\log(T/\text{s})$.
b Draw a graph of $\log T$ against $\log m$.
c Use the graph to determine:
   i   the value of n
   ii the value of c
d Write down the complete expression for T.

Logarithms to different bases

So far we have assumed that logarithms are taken to base 10. The word 'base' refers to the number that we are raising to a given power. For example, in $\log_{10} 1000 = 3$ the suffix 10 after log refers to the base being used. When it is not specified, it is generally assumed to be base 10. The button marked 'log' or sometimes 'lg' on your calculator is for calculating logs to the base 10.

However, logarithms can be taken to any base and other bases are used. In A-level Physics, another base that is used is the 'natural number', e, which is 2.718. It was introduced in Unit 7. Logs to the base e are usually referred to as natural logs. Because they are so widely used, your calculator will probably have a button especially for calculating logs to the base e. This button may be labelled 'ln'.

The principle is exactly the same as for logs to the base 10. If $x = e^y$, then $\log_e x = y$ so that finding $\log_e x$ and e^x are exactly reciprocal mathematical operations.

As an example, in Unit 7 we used the expression for the half-life, $t_{\frac{1}{2}}$, of a radioactive material, which is $t_{\frac{1}{2}} = \frac{\log_e 2}{\lambda}$, where λ is the decay constant or probability per unit time that a nucleus will decay. This expression is commonly written $t_{\frac{1}{2}} = \frac{\ln 2}{\lambda}$, where ln is taken to indicate a logarithm to base e.

Depending on your course or the options you take at A-level, you may also use logs to the base 2. These are used in electronics and communications. Digital systems use binary counting, or base 2, and each digit of a binary number is called a bit, which is short for binary digit. A binary digit can be 0 or 1 only.

Two binary digits can have four, or 2^2, different combinations: 00, 01, 10 and 11. Three binary digits can have 8, or 2^3, different combinations, and so on. To have 1024 different combinations requires $\log_2 (1024) = 10$ bits. Thus logs to the base 2 are used to calculate the number of bits required to encode a certain amount of information. As an example, a computer monitor is able to display more than 16 million colours. Taking the logarithm to the base 2 of this number gives $\log_2 (16 \times 10^6) = 23.9$, so 24 bits are needed to encode all these different colours. Early computers were only able to generate 256 colours, which requires just 8 bits.

A Worked examples

Use logs to the base e to calculate the following:

a 4640×2320

Step 1: use the ln button of the calculator to calculate the logs.

$\ln 4640 = 8.4425$ and $\ln 2320 = 7.7493$

Step 2: add the two logs.

$8.4425 + 7.7493 = 16.1918$

Step 3: use the e^x button on the calculator to find the result.

$e^{16.1918} = 1.076 \times 10^7$

(As before, doing the calculation normally on the calculator also gives 1.076×10^7.)

Full worked solutions at www.hoddereducation.co.uk/essentialmathsanswers

b $\log_2(256)$

Step 1: on your calculator press the key that shows $\log_\square \square$ or similar.

Step 2: enter '2' in the first box.

Step 3: use the cursor button to move the cursor to the second box and enter '256' and press the equals button. Your calculator will display 8.

c **The equation for radioactive decay is $N = N_0 e^{-\lambda t}$. Take logs to the base e (ln) to show that this equation can take the form of the equation for a straight line, $y = mx + c$.**

Step 1: take logs to base e of $N = N_0 e^{-\lambda t}$.

From the definition of logarithms we know $\ln e^{-\lambda t} = -\lambda t$.

Therefore, taking logs, $\ln N = \ln N_0 - \lambda t$.

Step 2: re-arrange the equation.

$\ln N = -\lambda t + \ln N_0$

Step 3: compare this with $y = mx + c$ to decide which terms correspond to x, to y, to the gradient, m, and to the intercept, c.

Here the intercept on the y-axis is $\ln N_0$ and the gradient of the graph is $-\lambda$.

B Guided questions

Copy out the workings and complete the answers on a separate piece of paper.

1 Find the natural log of 2.0.

Step 1: on your calculator press the 'ln' button. The display will probably show 'ln('.

Step 2: enter '2.0)' and press the equals button.

2 $e^x = 6.05$. Calculate x.

To find x, find the natural log of 6.05.

Step 1: on your calculator press the 'ln' button.

Step 2: enter '6.05)' and press the equals button.

3 $x = 3.5$. Calculate e^x.

Step 1: on your calculator, select the e^x function which is probably 'shift' + 'ln'. Your display will show e^x.

Step 2: enter '3.5' and press the equals button.

4 $x = -2.0$. Calculate e^x.

Step 1: on your calculator, select the e^x function. Your display will show e^x.

Step 2: enter '−2' and press the equals button.

5 The equation $N = N_0 e^{-\lambda t}$ gives the relationship between the number of atoms, N, in an initial sample, N_0, that remain to decay after a time, t, if the decay constant is λ.

A sample of cobalt-60 initially has 6.0×10^{24} atoms.

Calculate the number left to decay after 8.0 years.

$\lambda = 0.132\,\text{y}^{-1}$

Step 1: use $N = N_0 e^{-\lambda t}$ and substitute all the given values.

Since the time and λ are both given in years, you do not need to convert to seconds since the index for e must have no unit.

Step 2: on your calculator, select the e^x function and enter '-0.132×8.0' (be careful not to forget the '−') followed by the equals button.

Step 3: on most calculators you can then multiply the value obtained directly by 6.0×10^{24} to give your answer.

6 The relationship between the charge, Q, on a capacitor of value C, discharging through a resistor of value R, with time, t, is $Q = Q_0 e^{-\frac{t}{RC}}$, where Q_0 is the charge at time $t = 0$.

By taking natural logs of the equation, put it into the form of an equation for a straight line and explain the significance of the gradient and the intercept.

Step 1: take logs to base e of $Q = Q_0 e^{-\frac{t}{RC}}$

Remember that $\ln(e^x) = x$, and that, in this case, x is negative. When two quantities are multiplied, their logs are added.

Therefore, taking logs, $\ln Q =$ _____

Step 2: re-arrange the equation.

Step 3: compare the equation with $y = mx + c$ to decide which terms correspond to x, to y, to the gradient, m, and to the intercept, c.

Do not forget to take into account any negative signs when explaining the significance of the terms.

C Practice questions

7 Calculate the following:
   a $10^{5.3}$
   b $e^{1.6 \times 10^{-3}}$
   c 2^{24}
   d $\log_{10} 24$
   e $\ln 2$

8 In the equation, $t_{\frac{1}{2}} = \frac{\ln 2}{\lambda}$, $\lambda = 9.9 \times 10^{-3}$.

   Calculate $t_{\frac{1}{2}}$.

9 The potential difference, V, at time, t, across a capacitor of value C that is discharging through a resistor of value R is given by $V = V_0 e^{-\frac{t}{RC}}$. In an experiment to measure the potential difference, the graph in Figure 8.7 is obtained.
   a Calculate the initial potential difference.
   b Calculate the value of RC.
   c If R is $100\,\text{k}\Omega$, find C.

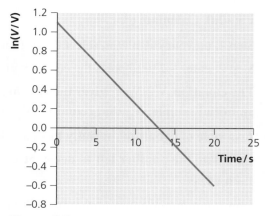

Figure 8.7

9 Uncertainty

Calculating uncertainty

Whenever you make measurements, there will be uncertainty in the value obtained. The more precise the measurements, the smaller the uncertainty. That does not mean your results are necessarily more accurate. Accuracy is how close your results are to the true or accepted value of a quantity. Precision is how close your results are to each other. In your A-level Physics course you will deal with ways of increasing both precision and accuracy in experimental results. The purpose of this section is to explain the mathematics of how to combine uncertainties when calculating results.

Suppose you measured the period of a pendulum and obtained the following results:

Table 9.1

Measurement	1	2	3	4	5	6	7	8	9	10
Time/s	2.2	2.1	2.2	2.3	2.1	2.2	2.4	2.3	2.1	2.2

To estimate the period you find an average or mean value for the time, T, which is 2.21 s (see Unit 4).

However, when dealing with experimental results, it is important to know the degree of uncertainty in your results. There are a number of ways of estimating the uncertainty in a set of results, sometimes referred to as confidence limits or standard error. The way that we use here, as appropriate for A-level Physics, is to calculate the range and spread of your results.

Range

The range of your results is the difference between the maximum value and the minimum value. In the case of the measurements of the periodic time for a pendulum, given in Table 9.1, this is $2.4 - 2.1 = 0.3$ s.

Spread

The spread is simply $\pm$ half the range. Therefore, for the pendulum, the spread of the measured periodic times is $\Delta T = \pm 0.15$ s. This is a measure of the uncertainty in the results. It is certainly a 'worst case' value since it gives the range in which 100% of the results obtained lie. This method does not take into account the benefit of taking more measurements of a value, which would decrease the uncertainty, but more sophisticated methods of calculation are not required for A-level.

Uncertainty

The result and the uncertainty in an experimental result, in this case $2.21 \pm 0.15\,\text{s}$, may be expressed as a fraction or a percentage. In this case, the fractional uncertainty is $\frac{\Delta T}{T}$ and the percentage uncertainty is $\frac{\Delta T}{T} \times 100\%$.

In general, the fractional uncertainty in a quantity $x = \frac{\Delta x}{x}$ and the percentage uncertainty in a quantity $y = \frac{\Delta y}{y} \times 100\%$.

Returning to the example of the periodic time of a pendulum, the fractional uncertainty is $\frac{\Delta T}{T} = \frac{0.15}{2.21} = 0.068$. Since the uncertainty is expressed to only one significant figure, this would be quoted as 0.07. The percentage uncertainty is therefore 7%.

We do not need to have a range of results in order to calculate a fractional or percentage uncertainty. For example, we might have measured the length, l, as 1212 mm but, given the difficulty of measuring to the centre of the pendulum bob and the uncertainty in reading a ruler, we might reasonably estimate our uncertainty in this measurement, Δl, as ± 4 mm.

This gives a fractional uncertainty of $\frac{\Delta l}{l} = \frac{4}{1212} = 3.3 \times 10^{-3}$ or 0.3%.

Uncertainties in quantities that are multiplied or divided

Many equations have quantities multiplied by each other and we need to know how uncertainties in each of these should be combined in the multiplication. For example, if we wanted to use our values of T and l in the example of the pendulum to calculate a value for the gravitational field strength, g, we would use the equation $T = 2\pi\sqrt{\frac{l}{g}}$ which, when re-arranged to make g the subject (see Unit 4), gives $g = \frac{4\pi^2 l}{T^2}$. We need to know how to combine the uncertainty of 7% in T with the uncertainty of 0.3% in l to give an uncertainty in g.

In general, if we have an equation $w = \frac{x^a \times y^b}{z^c}$ (or $w = x^a \times y^b \times z^{-c}$) then the fractional uncertainty in w, $\frac{\Delta w}{w}$, is given by $\frac{\Delta w}{w} = a\frac{\Delta x}{x} + b\frac{\Delta y}{y} + c\frac{\Delta z}{z}$. Any constants, such as $\frac{1}{2}$ or π, are ignored because there is no uncertainty in their values. Notice that even though the sign of c is negative, we still add $c\frac{\Delta z}{z}$.

In other words, to calculate the overall fractional uncertainty for quantities that are multiplied or divided, you add the fractional uncertainties multiplied by their index. The same would be true for percentage uncertainties.

This may seem surprising and perhaps counter-intuitive but there is a mathematical proof of this process, which you do not need and is not covered here.

Taking our example of the pendulum, we are using the equation $g = \frac{4\pi^2 l}{T^2}$.

Therefore the fractional uncertainty in g, $\frac{\Delta g}{g}$ is given by $\frac{\Delta l}{l} + 2 \times \frac{\Delta T}{T}$. This is $3.3 \times 10^{-3} + 2 \times 0.068 = 0.14$, which is 14%. The constant $4\pi^2$ has no uncertainty, so it is ignored. Notice that the larger percentage uncertainty in T compared with the percentage uncertainty in l makes the dominant contribution to the overall uncertainty in g and so it would be far more useful to take steps to reduce the uncertainty in T than the uncertainty in l.

We then calculate g using our mean values as $g = \frac{4\pi^2 \times 1.212}{2.21^2} = 9.80\,\text{N}\,\text{kg}^{-1}$ and calculate the actual uncertainty since $\frac{\Delta g}{g} = 0.14 \Rightarrow \Delta g = 0.14 \times 9.8 = 1.4\,\text{N}\,\text{kg}^{-1}$.

We then quote our experimental results as $g = 9.8 \pm 1.4\,\text{N}\,\text{kg}^{-1}$.

Our uncertainty should include the 'true' or accepted result and, in this case, it does.

Uncertainties in quantities that are added or subtracted

Although less commonly encountered in A-level Physics, it is sometimes necessary to combine uncertainties in quantities that are added or subtracted. This is done in a different way from the way when the quantities are multiplied or divided.

In general, if we have an equation $w = x + y - z$ then $\Delta w = \Delta x + \Delta y + \Delta z$ (note the + sign before Δz). In other words, rather than adding the fractional or percentage uncertainty, the actual uncertainties are added.

Take, for example, the equation $R_T = R_1 + R_2$. With a suitable resistance meter we might be able to measure the resistance of our resistors R_1 and R_2 to $\pm 2\%$.

Measured values are:

$R_1 = 120\,\Omega$ with ΔR_1 being $120 \times 0.02 = \pm 2\,\Omega$

$R_2 = 150\,\Omega$ with ΔR_2 being $150 \times 0.02 = \pm 3\,\Omega$

To calculate the uncertainty in R_T we add ΔR_1 and ΔR_2 to give $2 + 3 = \pm 5\,\Omega$.

Therefore $R_T = R_1 + R_2 = 120 + 150 = 270 \pm 5\,\Omega$.

The overall percentage uncertainty in RT is $\frac{5}{270} \times 100\% = 2\%$.

A more complicated situation arises if we have an equation that includes quantities that are both added and multiplied. For example, suppose we have the equation $s = ut + \frac{1}{2}at^2$ and the values of the quantities and the uncertainties are as follows:

$u = 5.0 \pm 0.1\,\text{m}\,\text{s}^{-1}$

$t = 50 \pm 1\,\text{s}$

$a = 1.5 \pm 0.06\,\text{m}\,\text{s}^{-2}$

We want to calculate a value for s including the uncertainty in s, Δs.

Step 1: calculate the fractional uncertainty in each quantity.

$\frac{\Delta u}{u} = \frac{0.1}{5} = 0.02$

$\frac{\Delta t}{t} = \frac{1}{50} = 0.02$

$\frac{\Delta a}{a} = \frac{0.06}{1.5} = 0.04$

Step 2: calculate the value of each term of the equation.

$$ut = 5 \times 50 = 250 \, \text{m}$$

$$\frac{1}{2}at^2 = \frac{1}{2} \times 1.5 \times 50^2 = 1875 \, \text{m}$$

Step 3: calculate the value of s.

$$s = 250 + 1875 = 2125 \, \text{m}$$

Since the data is given to two significant figures, $s = 2.1 \times 10^3 \, \text{m}$.

Step 4: calculate the uncertainty in each term.

$$\frac{\Delta(ut)}{ut} = \frac{\Delta u}{u} + \frac{\Delta t}{t} = 0.02 + 0.02 = 0.04$$

This means that $\Delta(ut) = (ut) \times 0.04 = 250 \times 0.04 = 10 \, \text{m}$

$$\frac{\Delta(\frac{1}{2}at^2)}{\frac{1}{2}at^2} = \frac{\Delta a}{a} + 2 \times \frac{\Delta t}{t} = 0.04 + 2 \times 0.02 = 0.08$$

Remember that, in calculating fractional uncertainty, any constants are ignored.

This means that $\Delta(\frac{1}{2}at^2) = (\frac{1}{2}at^2) \times 0.08 = 1875 \times 0.08 = 150 \, \text{m}$.

Step 5: calculate the total uncertainty in s.

$$\Delta s = \Delta(ut) + \Delta(\tfrac{1}{2}at^2) = 10 + 150 = 160 \, \text{m}$$

Again, since we are given the uncertainty data to one significant figure, we write

$$\Delta s = 2 \times 10^2 \, \text{m}.$$

Step 6: write the value of s, including its uncertainty.

$$s = 2.1 \times 10^3 \pm 2 \times 10^2 \, \text{m}$$

which is a fractional uncertainty of $\frac{\Delta s}{s} = \frac{2 \times 10^2}{2.1 \times 10^3} = 0.1$ or 10%.

Fortunately, at A-level, there are few situations where you might be required to complete this process but, by following the six steps above, you will be able to tackle any that do arise.

A Worked examples

a The wavelength of light diffracted at an angle θ by a diffraction grating is given by $n\lambda = d\sin\theta$. In an experiment, a diffraction grating with a slit separation, d, of 3.3×10^{-6} m is used and the angle of the first-order diffraction pattern of some yellow light is found to be $10.2°$. The uncertainty in the value of d is $\pm 0.1 \times 10^{-6}$ m and the uncertainty in measuring the angle is $\pm 0.1°$.

Calculate the wavelength of the yellow light and the uncertainty.

Step 1: use the given values to calculate λ.

For the first-order diffraction pattern, $n = 1$, and we are given that $d = 3.3 \times 10^{-6}$ and $\theta = 10.2°$, so

$$\lambda = 3.3 \times 10^{-6} \times \sin(10.2) = 5.8 \times 10^{-7}\,\text{m}$$

Step 2: calculate the uncertainty in d.

$$\frac{\Delta d}{d} = \frac{0.1 \times 10^{-6}}{3.3 \times 10^{-6}} = 0.03$$

Step 3: calculate $\Delta \sin\theta$ by calculating the range of $\sin\theta$ and then dividing by 2 to get the spread (see earlier text).

$$\Delta \sin\theta = \frac{\sin(10.3) - \sin(10.1)}{2} = 1.72 \times 10^{-3}$$

Step 4: calculate the uncertainty in $\sin\theta$.

$$\frac{\Delta \sin\theta}{\sin\theta} = \frac{\Delta \sin\theta}{\sin(10.2)} = \frac{1.72 \times 10^{-3}}{0.1771} = 0.01$$

Step 5: calculate the uncertainty in λ.

$$\frac{\Delta \lambda}{\lambda} = \frac{\Delta d}{d} + \frac{\Delta \sin\theta}{\sin\theta} = 0.03 + 0.01 = 0.04$$

Step 6: calculate $\Delta \lambda$.

$$\Delta \lambda = \lambda \times 0.04 = 5.8 \times 10^{-7} \times 0.04 = 0.2 \times 10^{-7}\,\text{m}$$

Step 7: write the value of λ, including its uncertainty.

$$\lambda \pm \Delta \lambda = 5.8 \times 10^{-7} \pm 0.2 \times 10^{-7}\,\text{m}$$

Ideally this should include the accepted or true value which, in this case, is 5.89×10^{-7} m.

b The formula for the total resistance, R_T, of two resistors, R_1 and R_2, in series is given by $R_T = R_1 + R_2$.

A resistance meter measures R_1 to be $1500 \pm 10\,\Omega$ and R_2 to be $3250 \pm 20\,\Omega$.

Calculate the value of R_T, including the uncertainty.

Step 1: use the given values to calculate $R_T = R_1 + R_2$.

$$R_T = 1500 + 3250 = 4750\,\Omega$$

Step 2: add the known uncertainties.

$$\Delta R_T = \Delta R_1 + \Delta R_2 = 10 + 20 = 30\,\Omega$$

Step 3: give the final answer as $R_T \pm \Delta R_T$.

$$R_T = 4750 \pm 30\,\Omega$$

B Guided questions

Copy out the workings and complete the answers on a separate piece of paper.

1 In an experiment you drop a small ball through a distance, s, of 1.5 m and measure the time, t, taken for it to hit the ground. The uncertainty in the measurement of the distance fallen is ±0.5 cm.

The measured times are as follows:

Table 9.2

Measurement	1	2	3	4	5	6	7	8	9	10
Time / s	0.56	0.56	0.54	0.56	0.56	0.56	0.58	0.53	0.53	0.55

a **Calculate the average value for the time from Table 9.2.**

Step 1: add all the values together and divide by 10.

Step 2: keep your answer to one more significant figure than the times given.

b **Calculate the range and spread of the results.**

Step 1: subtract the minimum value from the maximum value.

range = 0.58 s − 0.53 s

Step 2: find half the range to give the spread, Δt. Remember to put ± before this value.

$\Delta t =$ _____

c **Calculate the fractional and percentage uncertainty in the time.**

- The fractional uncertainty is $\frac{\Delta t}{t}$.
- The fractional uncertainty has no unit since both Δt and t have the same unit, s.

d **Calculate the fractional and percentage uncertainty in the value of the distance fallen.**

Step 1: check the data in the question and convert to the same units if necessary.

Δs is 0.5 cm and s is 1.5 m. These must be in the same units so it is probably easiest to take 1.5 m as 150 cm.

Step 2: calculate $\frac{\Delta s}{s}$.

e **The acceleration of the ball is given by $a = \frac{2s}{t^2}$.**

Calculate a value for a.

Step 1: substitute the value for s given in the question and your value for the average time, t, into the equation.

Step 2: calculate the value and remember to put the correct unit for acceleration after your answer.

f Calculate a value for the uncertainty in *a*.

The rule for finding the total fractional uncertainty when quantities are multiplied or divided is $\frac{\Delta a}{a} = \frac{\Delta s}{s} + 2 \times \frac{\Delta t}{t}$.

Step 1: substitute your values and calculate.

Step 2: to find a percentage uncertainty from a fractional uncertainty multiply by 100%.

Step 3: once you have the fractional uncertainty, $\frac{\Delta a}{a}$, you can find the actual uncertainty, Δa, by multiplying the value calculated in part **e** by the fractional uncertainty.

g Quote your final value for *a* including the uncertainty.

Step 1: write this using your answer to part **e** ± your answer to part **f**.

Step 2: add the correct unit.

2 The total capacitance, C_T, of two capacitors, C_1 and C_2, in parallel is given by $C_T = C_1 + C_2$.

A capacitance meter measures $C_1 = 4700\,\mu F \pm 230\,\mu F$ and $C_2 = 2200\,\mu F \pm 110\,\mu F$.

Calculate the total capacitance of the combination including the uncertainty.

Step 1: use the given values to calculate $C_T = C_1 + C_2$

$C_T = 4700\,\mu F + 2200\,\mu F =$ _____

Step 2: add the two known uncertainties.

$\Delta C_T = 230 + 110 = 340\,\mu F$

Step 3: give the final answer as $C_T \pm \Delta C_T$.

Remember to put the correct unit at the end.

C Practice questions

3 In an experiment to measure the speed of sound, the time, *t*, taken for a pulse of sound to travel a distance, *s*, of 200 m ± 1.0 m is measured 10 times.

The following measurements were taken of the time, *t*.

Table 9.3

Measurement	1	2	3	4	5	6	7	8	9	10
Time / s	0.58	0.62	0.57	0.57	0.57	0.59	0.58	0.59	0.60	0.61

a Calculate the average time, *t*.
b Calculate the range and spread of the times.
c Calculate the fractional uncertainty in the time.
d Calculate the fractional uncertainty in the distance, *s*.
e Calculate a value for the speed of sound including the uncertainty.

4 A voltmeter and an ammeter are used to measure the resistance of a resistor.

The reading on the voltmeter is 4.30 V ± 0.050 V and the reading on the ammeter is 0.50 A ± 0.050 A.

Calculate the value of the resistance and the uncertainty.

5 The emf, ε, of a cell is given by $\varepsilon = V + Ir$, where V is the potential difference across a cell of internal resistance, r, when a current, I, is flowing in the circuit.

A cell is known to have an internal resistance of 0.120 ± 0.0010 Ω. A potential difference of 1.46 ± 0.0050 V is measured across the terminals when a current of 0.50 ± 0.050 A is flowing.

Calculate the emf, ε, of the cell and the uncertainty in the value.

Exam-style questions

1. At its furthest distance from Earth, Pluto is 7.5 billion km from Earth.
   a State the distance from Pluto to Earth in standard form in metres. (1)

   Radio waves travel at $3.0 \times 10^8 \, \text{m s}^{-1}$.
   b Calculate the time, in hours, taken for a radio signal transmitted from a space probe near Pluto to reach Earth. (2)

   Data from the probe can be transmitted and received on Earth at a rate of about 1.1×10^3 bits per second. A picture from the probe's cameras might require 2.5 Mbits of data.
   c Calculate the time, in minutes, taken to transmit one picture from Pluto to Earth. Give your answer to an appropriate number of significant figures. (2)
   d Estimate, to one significant figure, the number of pictures that can be received in a day's 8-hour observation time. (1)
   e Explain why data compression is important for communications with space probes. (2)

2. The refractive index of glass, $_{\text{air}}\mu_{\text{glass}}$, is equal to the ratio of the speed of light in air to the speed of light in glass.

   An experiment to measure the speed of light in glass fibre finds that a pulse of laser light takes $0.11 \, \mu\text{s}$ to travel 20 m through the fibre.
   a Calculate the speed of laser light in the glass fibre. (1)

   Light travels at $3.0 \times 10^8 \, \text{m s}^{-1}$ in air.
   b Calculate the ratio of the speed of light in air to the speed of light in glass. (1)
   c Write down the refractive index of glass. (1)

   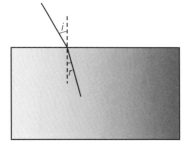

   Figure E.1

   When light is refracted as it enters a glass block, the ratio $\dfrac{\sin i}{\sin r} = \,_{\text{air}}\mu_{\text{glass}}$.
   d Assuming the glass block in Figure E.1 is made from the same glass as the glass fibre, calculate the angle, r, if angle $i = 30°$. (2)

3. An unstretched wire has a diameter of 0.24 mm.
   a Calculate its cross-sectional area. (2)

   The wire is now stretched by having a load attached to it. The diameter is remeasured and found to be 0.21 mm.

b Calculate the percentage change in the cross-sectional area of the wire. (2)

The wire was originally 2.0 m long and when it is stretched, its overall volume remains the same.

c Calculate the extension of the wire when loaded. (2)

The tensile strain in the wire is the ratio of the extension to the original length.

d Calculate the percentage tensile strain in the wire. (1)

4 The centripetal force required to keep a mass, m, orbiting at a speed, v, in a circle of radius, r, is given by $F = \frac{mv^2}{r}$.

   a Show that $F = \frac{4\pi^2 mr}{T^2}$ where T is the time for one complete orbit. (2)

The gravitational force between two masses, M and m, with their centres a distance r apart is given by $F = \frac{GMm}{r^2}$ where G is the universal gravitational constant.

   b By equating the two forces, show that $\frac{T^2}{r^3} = \frac{4\pi^2}{GM}$. (2)

The Moon orbits the Earth, which has a mass M, once every 28 days. The distance between the centre of the Earth and the centre of the Moon is about 3.8×10^5 km.

A geostationary satellite orbits the Earth once every day.

   c Calculate the ratio $\frac{T^2}{r^3}$ for any mass orbiting the Earth. (2)

   d Calculate the orbital radius of a geostationary satellite. (2)

5 A stone is thrown upwards from a cliff with a vertical speed of $5\,\text{m s}^{-1}$. The cliff is 100 m from the beach below. You can ignore air resistance.

The gravitational field strength, g, is $9.8\,\text{N kg}^{-1}$.

   a Calculate the time taken for the stone to land on the beach below the cliff. (4)

The horizontal speed of the stone is $2.0\,\text{m s}^{-1}$.

   b Calculate the horizontal distance from the thrower that the stone lands. (1)

6 Figure E.2 shows a bungee jumper. The bungee jumper has a mass of 60 kg and is attached to a cord of unstretched length 20 m and with a spring constant of $120\,\text{N m}^{-1}$. The gravitational field strength of the Earth is $9.8\,\text{N kg}^{-1}$.

   a Show that the change in gravitational potential energy, ΔE_p, from before the bungee jumper jumps to his lowest point is given by
   $\Delta E_\text{p} = (1.2 \times 10^4) + (5.9 \times 10^2 \times x)$ J. (2)

   b Show that the work done on the elastic in stretching it by a distance x is given by
   $\Delta W = 60x^2$ J. (2)

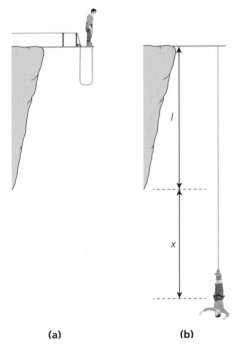

(a)　　　　　(b)

Figure E.2 A bungee jumper (a) before jumping and (b) at the lowest point when he is momentarily stopped

The change of gravitational potential energy, ΔE_p, is equal to the work done on the elastic, ΔW.

c Show that $60x^2 - 5.9 \times 10^2 x - 1.2 \times 10^4 = 0$. (2)
d Calculate x and hence the minimum height from which the bungee jumper must jump in order not to hit the ground. (4)

7 Boyle's law gives a relationship between the pressure, p, of an ideal gas and its volume, V, and states that pV = constant.
   a Re-arrange the equation to make p the subject. (1)
   b i State the quantities that should be plotted on the x-axis and the y-axis of a graph in order to obtain a straight line. (1)
      ii State what is represented by the gradient of the straight line. (1)

A student carries out the experiment and obtains the following results for p and V.

Table E.1

Volume / cm³	Pressure / kPa
41	100
35	120
30	140
26	160
24	180
21	200
19	220
18	240
16	260
15	280

c i Plot a suitable straight-line graph of these results. (3)
   ii Measure the gradient of the graph and explain its significance. (2)

8 Faraday's law states that the magnitude of the emf, ε, induced in a coil is the rate of change of flux linkage, $\frac{\Delta(N\varphi)}{\Delta t}$. Because of Lenz's law, $\varepsilon = -\frac{\Delta(N\varphi)}{\Delta t}$.

In a generator, the flux linkage for a rotating coil varies as shown in Figure E.3.

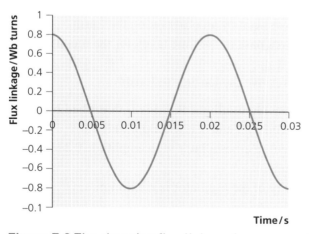

Figure E.3 The changing flux linkage for a generator coil

a Explain why the maximum emf occurs at times 0.005 s, 0.015 s, 0.025 s, etc. (2)
b Calculate the maximum emf generated by the rotating coil. (3)

9 A rubber band does not obey Hooke's law when stretched but stretches as shown in Figure E.4.

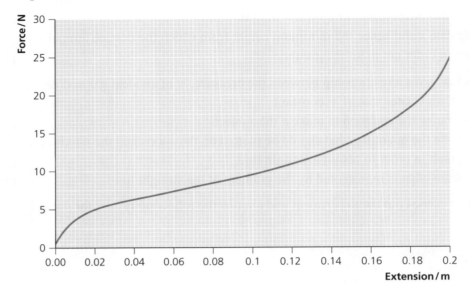

Figure E.4 The extension of a rubber band

a Indicate on a copy of the graph the area that corresponds to the work done in extending the rubber band by 20 cm. **(1)**
b Calculate the work done extending the band by 20 cm. **(3)**

10 Special relativity gives an expression for the time, t, measured on a clock moving at a relative speed, v, to a clock which measures a time, t_0.

$t = \gamma t_0$, where γ is the time dilation factor, $\gamma = \dfrac{1}{\sqrt{1 - v^2/c^2}}$

c is the speed of light, which is $3.00 \times 10^8 \,\mathrm{m\,s^{-1}}$.
A muon is created in the upper atmosphere about 60 km above the Earth. It travels at $2.98 \times 10^8 \,\mathrm{m\,s^{-1}}$ towards the Earth.

a Calculate a value for the time dilation factor, γ, for the muon. **(2)**
b Calculate the time taken for the muon to travel 60 km. **(1)**

The half-life of a muon is about 2 μs.

c Calculate the half-life of a muon as seen by an observer on the Earth. **(2)**
d Show that about 12 muon half-lives pass in the time taken for the muon to reach the Earth. **(1)**
e What proportion of muons will survive the journey to Earth? **(1)**

11 The displacement, s, of a simple harmonic oscillator at time t is described by the equation $s = A\cos\omega t$, where ω is a constant and A is the amplitude.

a Sketch a graph of the displacement of an oscillator with amplitude 0.2 m and $\omega = 4\pi$ for the first second of motion. **(2)**
b Label the points where the velocity of the oscillator is maximum. **(2)**

The acceleration of the oscillator at time t is given by $a = -A\omega^2 \cos\omega t$.

c Calculate the acceleration of the oscillator at time $t = 0.3$ s. **(2)**
d Calculate the maximum acceleration of the oscillator. **(1)**

12 When light is viewed through a diffraction grating with slit separation d, the angle θ at which light of wavelength λ is seen is given by $n\lambda = d\sin\theta$, where n is an integer.

A particular diffraction grating has 800 lines mm^{-1} and is used to look at a narrow monochromatic light source.
  a Calculate the separation of the lines on the diffraction grating, in m. (1)

The angle at which the first-order image appears ($n = 1$) is 23.6°.
  b Calculate the wavelength of the light. (2)
  c Calculate the angle at which the second-order image will be seen ($n = 2$). (2)
  d Explain whether you would expect to see a third-order image ($n = 3$). (2)

13 Figure E.5 shows a plane flying north at 200 m s^{-1} relative to the air with a cross-wind of 30 m s^{-1} relative to the ground towards the east.
  a Calculate the magnitude of the resultant velocity of the plane relative to the ground. (2)
  b Calculate the angle the resultant velocity makes to north. (2)

The pilot wants to fly due north at 250 m s^{-1} with the same cross-wind.
  c Calculate the velocity with which she must fly relative to the air. (3)

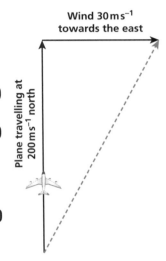

Figure E.5

14 An experiment to measure the force between two charges is shown in Figure E.6.

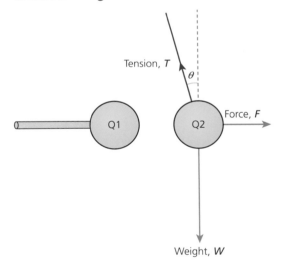

Figure E.6

The two light spheres are charged with equal charges.

Sphere Q1 is held on an insulating rod and sphere Q2 is suspended by a silk thread.

The spheres each have a mass of 1.0 mg.

In the position shown, the thread is at an angle of 15° to the vertical.

The gravitational field strength, g, is 9.8 N kg^{-1}.
  a Calculate the tension in the silk thread. (2)
  b Calculate the force between the two spheres. (2)

The separation of the two spheres is 9.0 cm and the charge on each sphere is measured to be 1.5 nC.

The force between the spheres is given by Coulomb's law, $F = \frac{kQ_1Q_2}{r^2}$.

   c Calculate a value for k. **(2)**

15 A crane lifts a load of 8.0×10^3 kg with a steel rope of diameter 16 mm.

The rope is 30 m long and has a Young modulus of 2.0×10^{11} Pa.

The Young modulus, E, is defined as

$$E = \frac{\text{tensile stress}}{\text{tensile strain}} = \frac{F/A}{e/l}$$

where F is the tensile force, A is the cross-sectional area, e is the extension and l is the original length.

The gravitational field strength, g, is 9.8 N kg^{-1}.

   a Calculate the tensile stress in the steel rope. **(3)**
   b Calculate the extension of the rope when lifting the load. **(2)**

The breaking stress of the steel rope is 7.4×10^8 Pa.

   c Calculate the maximum load that the crane can safely lift. **(3)**

16 A piece of Nichrome wire of length $l = 2.0$ m is connected into a circuit in order to measure its resistivity. The diameter of the wire is measured to be 0.32 mm.

The potential difference across the wire is 5.0 V and a current of 0.18 A flows through the wire.

   a Calculate the resistance of the wire, R. **(1)**
   b Calculate the cross-sectional area of the wire, A. **(1)**
   c Calculate the resistivity of the wire, ρ. **(2)**

The wire is now stretched to 2.1 m, keeping the overall volume the same.

   d Calculate the new resistance of the wire, R_2. **(2)**

17 The speedometer on a car measures the angular velocity of the wheels and calculates the linear speed from the measurement of angular velocity.

A car tyre has a diameter, when new, of 66 cm.

   a Show that when travelling at 30 mph (13.4 m s^{-1}), the speedometer is recording an angular velocity for the wheels of 41 rad s^{-1}. **(3)**

After many miles, the tyres are worn and the overall diameter of the tyre is now 65 cm.

   b Calculate the actual speed of the car when the speedometer reads 30 mph. **(1)**
   c Calculate the percentage difference between the measured speed and the actual speed. **(2)**

18 The ancient Greeks compared the diameter of the Moon and the diameter of the Earth and estimated that the diameter of the Earth is about 3.5 times that of the Moon. They also knew that the diameter of the Earth is about 13×10^6 m.

   a Calculate the approximate diameter of the Moon. **(1)**

An experiment to calculate the distance of the Moon from the Earth can be carried out by holding a small coin in front of your eye at full moon and moving it backwards and forwards until it just obscures the Moon, as shown in Figure E.7.

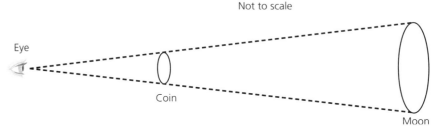

Figure E.7

In an experiment it is found that the coin has to be held 110 coin diameters away from the eye in order to just obscure the Moon.
b Estimate the distance to the Moon. (2)

In a solar eclipse, the Moon exactly covers the Sun. The diameter of the Sun is known to be 1.4×10^9 m.
c Estimate the distance from the Earth to the Sun. (2)

19 A radioactive source of strontium-90 has a half-life of 28.8 years. 5.0 g of pure, freshly prepared strontium-90 is obtained.

The Avogadro constant, $N_A = 6.02 \times 10^{23}$ mol^{-1}.
  a Calculate the decay constant for strontium. (2)
  b Calculate the activity of the sample of strontium-90. (3)
  c Calculate the activity of the sample after 50 years. (2)

20 A sample of iodine-131, used for medical treatment, is measured to have an activity of 4.5×10^9 Bq.
The half-life of iodine-131 is 8.02 days.
  a Calculate the decay constant for iodine-131. (1)
  b Calculate the number of atoms of iodine-131 present in the sample. (2)

The iodine-131 is administered to a patient. After 10 days the activity of the iodine is checked.
  c Calculate the expected activity after 10 days. (2)

In fact the measured activity is much less than calculated in part **c**.
  d Suggest a reason why the measured activity is lower than calculated. (1)

21 A capacitor is fully charged to 5 V and connected to a circuit containing two resistors as shown in Figure E.8.

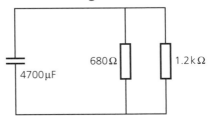

Figure E.8
  a Calculate the initial current in the circuit. (3)
  b Calculate the time constant RC. (1)
  c Calculate the potential difference across the capacitor after 1.5 s. (2)

22 A capacitor of value 1.0 µF is charged to a potential difference of 10 V.
  a Calculate the charge stored on the capacitor when it is charged. (1)

The charged capacitor is now connected directly to a second, uncharged, capacitor of value 10 µF.
  b Calculate the total value of the two capacitors in parallel. (1)

The charge on the first capacitor is now shared between the two capacitors.
  c Assuming that the total capacitance found in part **b** is charged with the total charge found in part **a**, show that the potential difference across the pair of capacitors is about 0.9 V. (1)
  d Calculate the charge on each of the capacitors separately. (2)

The work done charging a capacitor is given by $\Delta W = \frac{1}{2}QV = \frac{1}{2}CV^2$.

  e Calculate the work done when the 1.0 µF capacitor was charged to 10 V. (1)
  f Calculate the work done when:
    i   the 1.0 µF capacitor is charged to 0.9 V (1)
    ii the 10 µF capacitor is charged to 0.9 V (1)

23 If a mass of gas is compressed quickly so that no heat energy is lost, known as an adiabatic change, Boyle's law does not apply. Instead, pV^γ = constant.

In such a change, values of p and V are recorded.
  a By taking logs, write down this expression in a form that will allow a straight-line graph to be plotted. (2)

In an experiment, the following values of pressure and volume were recorded:

Table E.2

p/kPa	V/m³	log(V/m³)	log(p/Pa)
100	0.20		
109	0.19		
119	0.18		
131	0.17		
145	0.16		
162	0.15		
181	0.14		
205	0.13		
235	0.12		

  b Copy and complete the table and plot a graph that will allow you to find a value for γ and for the constant. (4)
  c Use your graph to calculate γ. (2)
  d Use your graph to find a value for the constant. (2)

24 In an experiment to measure the half-life of protactinium-234, a sample was prepared and the activity measured every 20 s.

The following results of corrected count rate against time were recorded.

Table E.3

Time/s	Corrected activity/Bq
0	160
20	131
40	108
60	88
80	72
100	59
120	49
140	40
160	33
180	27
200	22

The activity, A, of a radioactive source at time t is given by $A = A_0 e^{-\lambda t}$, where A_0 is the activity at time $t = 0$ and λ is the decay constant.

a Take natural logs of the equation to give an expression for a straight-line graph. **(1)**
b Plot a log graph of the results that will enable you to find a value for the half-life of protactinium. **(3)**
c Use your graph to find a value for the half-life of protactinium. **(2)**

25 In order to measure the periodic time of an oscillator, the periodic time is measured 10 times. The results are shown in the table below.

Table E.4

Measurement	1	2	3	4	5	6	7	8	9	10
Periodic time/s	5.7	5.5	5.3	5.5	5.6	5.7	5.4	5.6	5.3	5.4

a Calculate the mean of the periodic time. **(1)**
b Determine the uncertainty in the measurement. **(2)**
c Calculate the percentage uncertainty in the periodic time. **(1)**

The oscillator is a simple pendulum of length $7.5 \text{ m} \pm 0.01 \text{ m}$. It is being used to find a value for g, the gravitational field strength, where $g = \dfrac{4\pi^2 l}{T^2}$.

d Calculate a value for g and the uncertainty in the value. **(3)**

Appendix

Specification cross-reference

The numbers in each box refer to the section of the core parts of the relevant specification. Numbers in bold are second-year topics. However, some fundamental topics, for example, powers of 10, units, prefixes, standard form, calculations, re-arranging equations and evaluating equations, are relevant to almost every part of the course so the numbers refer to the first or most prominent sections only. The information in the table is intended as a guide only. You should refer to your specification for full details of the topics you need to know.

Topic in this book	AQA	OCR A	OCR B	Edexcel	WJEC/Eduqas	CCEA
1 Units, standard form and orders of magnitude						
Powers of 10	3.1	1.1 4.1	1.1	1.1	1.1	1.1
Units	3.1	2 3.1, 3.2	1.1, 2, 3.1	1.1	1.1	1.1
Prefixes	3.1	1.1, 2.1, 3.1, 3.2	1.1, 2, 3.1	1.1	1.1	1.1
Converting units	3.1	1.1, 2.1, 3.1, 3.2	1.1, 2, 3.1	1.1	1.1	1.1
Symbols used at A-level	3.1	3.2, 3.5	3.1, 4.2 **5.1, 5.2**	2.25, 2.27, 3.31 **10.162**	1.1, 2.5 **3.4, 4.3, 4.5**	1.1, 2.7
Standard form and significant figures	3.1	1.1, 3.2 4.1	1.1, 3.2, 4.1	1.1	1.1, 1.2, 1.3, 1.4, 1.6, 2.1, 2.2, 2.4, 2.5, 2.7, **3.5, 4.1, 4.2, 4.4**	1.1, 2.7
Calculations using standard form	3.1	1.1, 3.2	1.1, 3.2, 4.1	1.1	1.1, 1.2, 1.3, 1.4, 1.6, 2.1, 2.2, 2.4, 2.5, 2.7, **3.5, 4.1, 4.2, 4.4**	1.1
Orders of magnitude	3.1	3.1, 5.5, 6.4	1.1, 3.1, 3.2	1.2	1.5, 1.6, 2.1, 2.2, 2.5, 2.7 **3.1, 3.3, 3.4, 4.1, 4.2, 4.3**	1.1
2 Fractions, ratios and percentages						
Fractions	3.3	3.3 **6.3, 6.5**	2, 3.3, 3.2 **5.2**	3.31	1.1	1.8, 1.10 **5.2, 5.3**
Ratios	3.3 **3.7**	3.3 **6.3, 6.5**	3.1	2.30, 5.71	1.1, 1.4, 1.7, 2.6 **3.4**	2.2, 2.3 **5.5**
Percentages	3.4	3.3 **6.3, 6.5**	3.3	2.30	1.4 **3.4**	1.8

Topic in this book	AQA	OCR A	OCR B	Edexcel	WJEC/Eduqas	CCEA
3 Averages and probability						
Averages	3.1, 3.8	1.1	1.1, 2, 3.2	1.3	2.4	3.2
Probability	**3.8**	1.1, **6.4**	4.1, **5.1, 6.4**	**11.173**	2.7	**4.6**
4 Algebra						
Re-arranging equations	3.1, 3.3, 3.4, 3.5, **3.6, 3.7**	3.1, 3.2, 4.2, **5.3**	3.1, 3.2, 4.2, **5.1, 6.2**	2.9, 3.31, 3.37, 3.39, 4.54, **6.102**	1.1, 1.2, 1.5, 2.3, **4.3**	1.3, 1.9, **4.1, 4.5, 5.2, 5.3**
Evaluating equations	3.2, 3.3, **3.7**	3.1, 3.2, 3.4, 4.3, **5.4**	3.1, 3.2, 4.1, 4.2, **5.1, 6.1, 6.2**	2.9, 2.17, 3.37, 4.54, 5.68, **6.102, 6.104, 7.110, 8.134, 9.146**	1.2, 1.5, **4.3**	1.4, 1.6, 1.7, 1.10, 1.11, 1.12, 2.3, **4.1, 4.5, 5.2, 5.3**
Quadratic equations	3.4	3.1, **5.2**	3.1, 4.2, **5.1**	2.26, **6.102**	1.2, **3.1, 4.3**	1.5
5 Graphs						
Straight lines	3.1, 3.4	1.1, 3.1, 3.4	1.1, 3.2, 4.1, 4.2, **5.1, 5.2**	2.10, 5.68	1.5, 2.3, **4.4**	1.4, 1.11, 3.2, **4.1**
Shapes of graphs for different functions	3.5, **3.6, 3.7, 3.8**	3.4, 4.2, **5.3, 6.4**	3.1, 3.2, 4.2, **5.1, 5.2**	**7.129, 9.151, 10.156, 12.176, 13.183**	1.2, 1.6, 2.2, 2.4, **3.3**	1.10, **4.2**
Rates of change	3.4, 3.5, **3.7**	3.1, 3.5, **5.3, 6.1, 6.3, 6.4**	3.1, 4.2, **5.1**	2.11, **7.112**	1.2, 1.3, 2.1, **3.2, 4.2, 4.5**	1.5, **5.3, 5.5**
Area under a graph	3.4, **3.7**	3.1, **3.5, 6.1**	3.2, **5.1, 6.1, 6.2**	2.11, **7.112, 7.117**	1.2, 1.5, **3.4, 4.1**	1.8, **5.2, 5.3, 5.4**
6 Geometry and trigonometry						
Radians	3.3, **3.6, 3.7**	4.4, **5.2, 5.3**	4.1, **5.1, 5.2**	**6.103**	3.1	4.3
Degrees and radians	3.3, **3.6, 3.7**	4.4, **5.2, 5.3**	4.1, **5.1, 5.2**	**6.104**	3.1	**4.3, 5.5, 5.6, 5.7**
Sine, cosine and tangent	3.3, **3.6**	2.3, 3.1, 4.4	4.1, 4.2, **5.1**	2.14	1.1, 1.4, 2.5, 2.6, 3.2, **4.2**	2.2, 2.3, 2.4, **4.4, 5.5**
Small angles	**3.6**	4.4, **5.5**	4.1	5.84, **13.183**	1.4, 2.5, **4.2**	**4.4**
Pythagoras	3.4	2.3, 3.2	4.2	2.14	1.1, 2.5, 2.6, **4.2, 4.4**	1.2
Resolving vectors	3.4	2.3, 3.2	4.2	2.12, 2.13	1.1, **4.2, 4.4**	1.2
Areas and volumes of simple shapes	3.4, 3.5	3.2, **3.5**	3.1, 3.2	2.10, 3.39, 3.41	1.1, 1.5, 2.1, 2.2	1.10, **5.5**

Topic in this book	AQA	OCR A	OCR B	Edexcel	WJEC/Eduqas	CCEA
7 Exponential changes						
Radioactive decay	3.8	6.1	5.1	11.172, 11.173	3.5	4.6
Capacitor discharge	3.7	6.4	5.1	7.118, 7.120	4.1	5.4
8 Logarithms						
Understanding logarithms	3.8	6.1, 6.4, 6.5	5.1, 5.2	7.120, 11.173	3.5, 4.1, 4.2	4.6
Logarithmic scales	3.8	6.1, 6.4, 6.5	5.1, 5.2	7.120, 11.173	3.5, 4.1, 4.2	6.2
Using logarithms	3.8	6.1, 6.4, 6.5	5.1, 5.2	7.120, 11.173	3.5, 4.1, 4.2	4.6, 6.2
Logarithms to different bases	3.8	6.1, 6.4, 6.5	5.1, 5.2	7.120, 11.173	3.5, 4.1, 4.2	4.6
9 Uncertainty						
Calculating uncertainty	3.1	1.1, 2.2	1.1, 2	1.4	1.1, 1.5, 2.2, 2.3, 2.5, 2.6, 3.2, 4.1, 4.4	3.3
Uncertainties in quantities that are multiplied or divided	3.1, 3.5	1.1, 2.2	1.1, 2	1.4	1.1, 1.5, 2.2, 2.3, 2.5, 2.6, 3.2, 4.1, 4.4	6.2
Uncertainties in quantities that are added or subtracted	3.1	1.1, 2.2	1.1, 2	1.4	1.1, 1.5, 2.2, 2.3, 2.5, 2.6, 3.2, 4.1, 4.4	6.2